BIOLOGICAL STRUCTURE AND FUNCTION 9

SKIN

T0291511

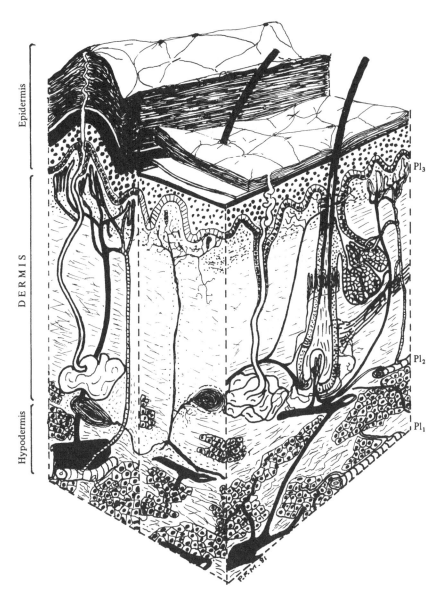

Diagrammatic representation of skin indicating both thick and thin skin (not to scale).
Pl $_{1-3}$, levels of the three plexuses of blood vessels.

BIOLOGICAL STRUCTURE AND FUNCTION

EDITORS

R. J. HARRISON
Professor of Anatomy
University of Cambridge

R. M. H. McMINN
Emeritus Professor of Anatomy
Royal College of Surgeons of England

SKIN

P. F. MILLINGTON

Reader in Bioengineering,
Head of Microscopy Division,
University of Strathclyde

R. WILKINSON

Lecturer in Bioengineering,
Bioengineering Unit,
University of Strathclyde

CAMBRIDGE UNIVERSITY PRESS

CAMBRIDGE

LONDON   NEW YORK   NEW ROCHELLE
MELBOURNE   SYDNEY

CAMBRIDGE UNIVERSITY PRESS
Cambridge, New York, Melbourne, Madrid, Cape Town, Singapore, São Paulo, Delhi

Cambridge University Press
The Edinburgh Building, Cambridge CB2 8RU, UK

Published in the United States of America by Cambridge University Press, New York

www.cambridge.org
Information on this title: www.cambridge.org/9780521106818

© Cambridge University Press 1983

First published 1983
This digitally printed version 2009

A catalogue record for this publication is available from the British Library

Library of Congress Catalogue Card Number: 82-14637

ISBN 978-0-521-24122-9 hardback
ISBN 978-0-521-10681-8 paperback

CONTENTS

PREFACE

To choose the title *Skin* is in one sense presumptuous, in another challenging. The subject is so vast that only a brief overview is possible in a short book. Indeed, a computer search of the Medlars database revealed over 28 000 publications on skin in the three years from 1978 to 1980. The majority of these occur in the fields of pathology, dermatology and, to a lesser extent, in cosmetics.

This book only deals with a limited range of topics and concentrates on those of particular interest to bioengineering, but even within these topics we have been circumspect in selecting descriptions of structure and behaviour to illustrate rather than provide a comprehensive account. Primarily, we describe normal human skin, but have sought illustrations from animal studies and pathological states where appropriate.

Inevitably, some of the mathematics used to describe the response of skin to mechanical forces may be unfamiliar to readers and a brief explanation is included in the text for those who have some mathematical background. The mathematical notation and symbols have been standardised, as far as possible, throughout the text, so many equations do not appear in the form presented in the original publications. To help the reader further, a list of symbols used will be found at the end of Chapter 4 and at the end of each section of Chapter 5, together with a summary of basic mathematical notation. Elsewhere, symbols are explained within the text.

We begin by considering the body surface – its appearance, structure and properties – but skin is a large organ which differs greatly in both structure and function from site to site. After considering skin in depth, a brief description of the site variation is given based essentially on structural parameters. The important physical and mechanical properties of skin are dealt with in separate chapters and the book is concluded by a brief survey of the response of skin to damage and ways in which healing may be assisted. It is perhaps appropriate that this final part also outlines the more recent attempts to provide substitutes for skin, for it is through this work that many of the functions of skin, which we take for granted, are highlighted.

Many people have assisted in the compilation of this book and we wish to thank particularly the Assistant Depute Librarian, Dr H. Cargill-Thomson, and other members of the Andersonian Library, University of Strathclyde, for their help in locating source material. We would also thank Mr H. Petrie (Andersonian Library) who assisted with the computer search of literature; the staff of the University of Glasgow Library who spent many hours locating papers and reference material through their stock rooms; Dr T. R. Steele, Department of Mathematics, University of Strathclyde, for his helpful discussions on tensor analysis; Professors R. M. H. McMinn (London) and R. J. Harrison (Cambridge) for their editorial suggestions and Professor J. P. Paul, Bioengineering Unit, University of Strathclyde, for his continued support; and Mrs P. Bissell for typing the manuscript. We are indebted to Dr A. M. Stoll, USA, and Dr R. P. Clark, CRC Bioengineering Division, Middlesex, UK, for kindly providing illustrations, and to Dr M. Ferguson-Pell and other colleagues of the Bioengineering Unit, University of Strathclyde, for their immediate response in providing material for incorporation in this book.

April 1982

1

THE BODY SURFACE

Skin is the organ which forms the boundary between the body and the external world. It helps to maintain internal homeostasis by thermoregulatory secretory activity, forms a self-repairing barrier against physical and chemical assault, provides an important route for the transmission of information about the external environment, and retains, in the layer immediately subjacent, a major food reserve. The appearance of skin, including hair and nails, provides both social signals and clues to the general state of well-being.

AREA

The skin comprises about one-twelfth of the body mass, yet the apparent surface area of the adult is only 1.2–2.0 m². This is due largely to the absence of major folds, plication and villi, and to the relatively smooth contours of the overall body surface. An approximate value for the surface area can be obtained from a knowledge of the height and body weight of an individual. A simple nomograph relating these variables is given in Fig. 1.

Calculation of the true body surface area is very complex and requires the summation of a large number of small measurements. Even then, assumptions relating to the creases and micro-folding of the stratum corneum need to be made. Hence attempts to measure total skin surface area are seldom made and it is questionable whether such measurements would have any useful applications in either dermatology or cosmetics.

TOPOLOGY

Skin surface patterns are formed by folding or ridge formation. In the early stages of development, the epidermis is formed from the ectodermal embryonic layer, and gives rise to hair follicles, apocrine, sebaceous and eccrine glands. The classical work of Pinkus (1910, 1927) forms the basis on which most later descriptions rest.

The early embryonic epidermis consists of a uniform sheet of cells two layers thick; the outer includes the periderm and epitrichial layer which completely covers the embryo, and the inner includes the stratum germinativum, (Fig. 2a). By 11 weeks, three layers are visible (the middle layer now being termed the stratum intermedium) (Fig. 2b). The basal layer now shows evidence of increased mitotic activity and the cells have become taller and

1

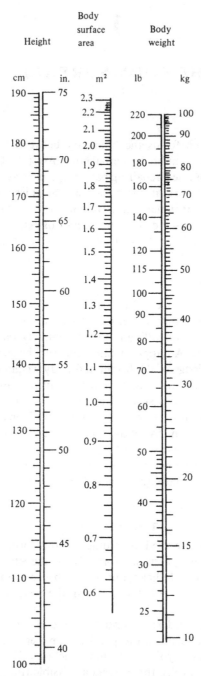

Fig. 1. Nomograph for determining body surface area from two variables: body weight and height. To use the nomograph, lay a straight edge to intercept height and weight at appropriate points. Read area from the intercept of the straight edge with the surface area line.

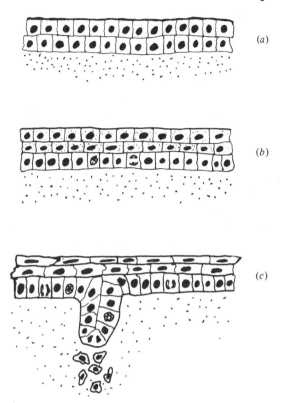

Fig. 2. Diagrammatic representation of embryonic growth of epidermis. (*a*) *Early embryo*. Two layers, the outer periderm and epitrichial layer, and an inner stratum germinativum. (*b*) *About 11 weeks*. Three layers; the middle layer is known as the stratum intermedium. (*c*) *About 12 weeks*. Basal layer becomes more columnar and begins to grow downwards. Below the downward growth, mesenchymal cells form focal aggregations. It is from these cells that hair follicles will develop.

more closely packed. The nuclei are rounder and larger. After a very short time, mesenchymal cells begin to aggregate below selected areas, some of which will develop later into hair follicles. Columnar basal cells begin to grow downwards (Fig. 2*c*), and a basement membrane develops at the epidermal–dermal junction. It is shortly after this that foetal ridges and lines are noted.

The first surface ridges are formed about the thirteenth week on the palmar and plantar surfaces of the tips of the digits and later extend over the entire volar surfaces of the hands and feet (Hale, 1952). The ridges lie over the dermal papillae through which the sweat ducts will penetrate to emerge through the elevated portions of the ridges. The patterns that develop at this time remain essentially the same throughout life, although the ridge width increases at the same rate as the growth of the hands and feet (Hale, 1949).

During the early period of growth, up to 14 weeks, the epidermal cells contain large quantities of glycogen, and the periderm displays microvilli. As the number of intracellular fibres increases, the glycogen diminishes, so that, by the sixteenth week, the periderm contains both fully keratinised and unkeratinised cells, with many subjacent cells containing keratohyalin. By the seventeenth week, the outer layer of cells is fully keratinised and exhibits only minimal glycogen content. The epidermis has all the essential structural characteristics of the adult.

Until the early 1970s there had been few attempts to describe systematically and in detail the surface characteristics of skin, other than the ridge patterns of the fingers, hands and feet. From both direct photography and impression or printing techniques of skin, descriptions were added to histologically derived information to build up a record of the body surface necessary for the medical needs of the time. Major developments in plastic surgery from 1937 to 1945, necessitated by the demand for urgent treatment of burns inflicted during World War 2, caused renewed interest in the cosmetic achievements of surgery. This in turn led to a number of attempts to add detail to the description of skin surfaces (Wolf, 1937, 1939; Hambrick & Blank, 1954; Oberste-Lehn, 1962; Moynahan & Engel, 1962; Sarkany, 1962; Facq, Kirk & Rebell, 1964).

The advent of scanning electron microscopy led to an increased interest in the surface structure of skin. However, the procedures used in the preparation of specimens had been based on those of conventional histology and caused large distortions primarily through shrinkage. A search for methods to give accurate topographical information has been a major concern since 1960.

Sarkany (1962) described a replication method for studying skin topography using a plastic material, but the method later introduced by Facq *et al.* (1964) has been the one on which most replica techniques have been based. They devised a system wherein a short length of acetone-soluble film with high adhesive properties was laid on a clean glass slide in a small pool of acetone and allowed to dry. Any slides with small bubbles showing Newtonian rings were discarded, as were any where the film edges were curled. Replicas were obtained by two variants of their basic technique. Flat resilient areas of the body (trunk or fleshy parts of the limbs), when held as near horizontal as possible, were flooded with acetone and the slide-borne film rolled onto the area so as to exclude all air between the film and skin. On highly contoured areas, such as the nose, digits or bony prominences, one or two drops of acetone were allowed to fall onto the film and the appropriate area of skin lowered onto the slide. In either case, contact was maintained until the film had dried, when it was carefully removed, starting at one corner and lifting evenly.

Various replica materials have been tried from time to time. The use of synthetic rubber was reported by Goldman *et al.* (1969) as being excellent for the relatively dry outer surface of skin. Kuokkanen (1972) used a silicone rubber paste, Xantopren (manufactured by Farbenfabriken Bayern, A.G., Germany), to obtain some excellent impressions of skin. She used a method similar to that of Facq. A small quantity of the paste was placed on a glass slide, and after mixing with a few drops of hardening substance, pressed onto the skin. When hard, the replica loosens readily from the skin but adheres to the slide. An alternative method employs Silcoset (ICI) which has been mixed with a fast-setting catalyst. A small quantity may be applied to the skin with a 'swab' stick which is rolled in a direction counter to that of the movement across the skin. Since all available catalysts can produce an unwelcome skin reaction, a small test area should be tried before undertaking multiple replicas or applying to a visible area of the skin. The replica can be lifted away as soon as set with the aid of a pair of forceps, holding it at one corner. Because Silcoset has good elastic properties, any small extension or distortion introduced during lifting, recovers almost immediately to give an excellent replica of skin.

A variation of the replica methods that is claimed to be an improved technique for studying the stratum corneum, was introduced by Marks & Dawber (1971). They replaced the plastic film and acetone of the earlier method by ethyl cyanoacrylate, which in the presence of small quantities of water or water vapour adheres very firmly to both skin and glass. A thin biopsy specimen of the stratum corneum can thus be obtained. Such a preparation may be stained before inspection by light microscopy to display not only the outer surface of the stratum corneum, but also the underlying surface of the removed corneocytes. Direct photographic enlargement of the adhered specimen permits the visualisation of major crease patterns.

Replicas and stripped specimens may be inspected directly and then photographed. After coating with a suitable conducting material, such as gold or gold–palladium, they may be mounted on a stub for inspection by scanning electron microscopy at magnifications up to × 50000.

First reports contained many illustrations of differences in skin patterns at different body sites, but most were concerned more with the techniques employed than with the site variations. Nevertheless, a general sequence of patterns was quickly established and forms the basis of the following descriptions.

Because of the need to shave away hair from sites of dense growth, skin patterns for scalp, pubic region, and mid-ventral thoracic regions in men are poorly documented. Except for the volar regions of the body, there appears to be a simple basic pattern to the creases in skin which is modified locally according to the direction and frequency of movement. In all major areas of the body the skin surface has a geometric pattern. The vertical direction of the crease is inwards towards the dermis, but the principal planar region is not flat, but curved outwards. These two features represent a built-in redundancy, which is essential for extensive movement of the skin surface

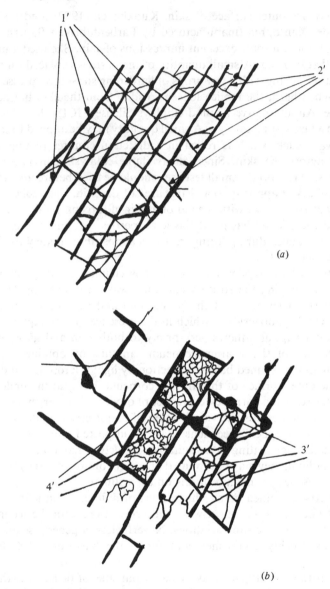

Fig. 3. Epidermal surface lines. (*a*) Primary (1′) and secondary (2′) lines. (*b*) The much smaller tertiary (3′) and quaternary (4′) lines. The quaternary lines coincide with cell outlines.

(Schellander & Headington, 1974). The linear creases cross each other to give rise to a lattice-type pattern. Although it has been reported that over limited areas these patterns are isometric, this is seldom the case. Generally the repetitive spacing in one direction is longer than in the other, and the commonest repeat figure is a rectangle or parallelogram. The rectangles are

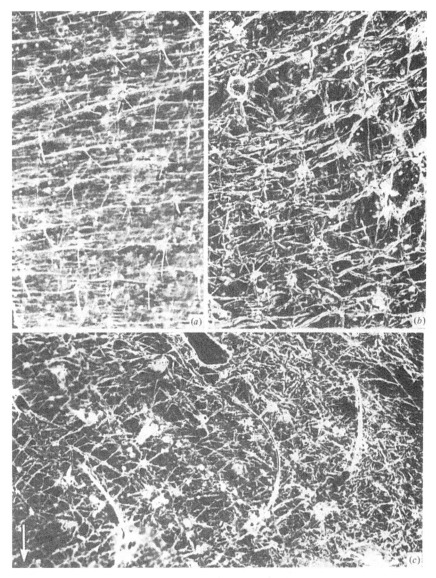

Fig. 4. Photographic enlargements (× 9.5) of crease patterns in stripped epidermal skin. (a) Forehead, (b) lower jaw, (c) neck. The arrow indicates the direction of the long axis of the body. Hairs are prominent in both jaw and neck specimens.

in turn crossed by one or two diagonals, which subdivide them into two or four triangles.

From low-magnification study, it would appear that the triangles are the smallest unit on many areas of skin, and these correspond to the primary and secondary lines proposed by Wolf (1939) and later subscribed to by Fujita

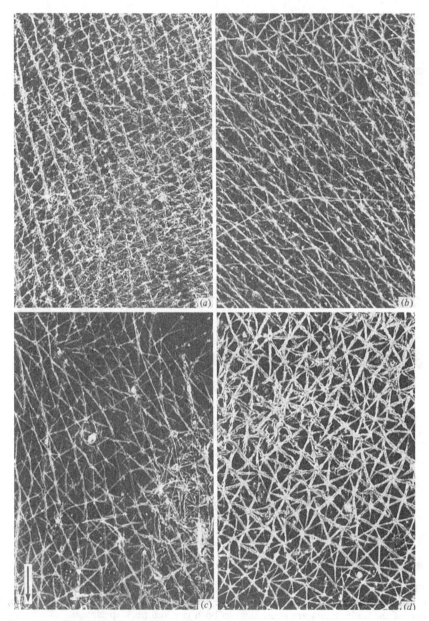

Fig. 5. Photographic enlargements (× 7.5) of crease patterns in stripped epidermal skin. (*a*) Thorax, (*b*) abdomen, (*c*) back, (*d*) buttock. The arrow indicates the direction of the long axis of the body.

(1973) & Hashimoto (1974) (Fig. 3*a*). At higher magnification, smaller creases can be detected along the edges of groups of corneocytes (tertiary lines), and the boundaries of individual corneocytes have been termed 'quaternary lines' (Fig. 3*b*).

Each triangle or sub-unit encloses a variable number of horny cells, which themselves are geometric, the most common shape being either hexagonal or pentagonal. The presence of the sub-units or tertiary lines is more common in sites with pronounced flexibility (wrist, knuckle, elbow, instep, etc.) and less common over the flatter surfaces of forearm, shin and trunk. The primary system of creases usually represents the macroscopically visible skin creases. The finding that this pattern prevails throughout the entire thickness of the horny layer can be demonstrated by successive strippings taken from the same site.

The *scalp* surface pattern would appear to be an unusual system of hexagonal lines, with finer lines tending to radiate from the centre, but with smaller hexagonal patterns occasionally visible (Schellander & Headington, 1974). By contrast, the *forehead* shows a very rudimentary rectilinear pattern (Fig. 4*a*). Orientation of the primary creases is in line with the major vertical and horizontal body axes. Secondary creases radiate from certain cross-over points which contain hair follicles. These creases seldom extend beyond the first primary crease. The *cheek* presents a pattern dominated by crease lines radiating from the insertion points of hair follicles. The lower jaw, too, is dominated by radiating creases, but the rectilinear creases can be readily distinguished (Fig. 4*b*).

The *neck* pattern is dominated by the circumferential creases, which appear to bifurcate and anastomose to form an interlinked network, (Fig. 4*c*). By scanning electron microscopy the same patterns appear to have numerous short cross-creases.

The *thorax, abdomen, back* and *buttock* all have very similar surface patterns. The dimensions of the patterns vary from site to site and are distorted wherever the skin is relaxed (Fig. 5*a–d*). The basic pattern described above dominates the normal appearance and in all cases the two primary creases are inclined to the vertical axis of the body so that they appear as a pattern of diamonds, interlaced by secondary creases. In many areas the apparent pattern is one of triangles. The length of the sides of the triangles varies from about 0.5 mm to 0.9 mm.

The *calf* has a pattern which is principally one of simple parallelograms, with sides inclined at an angle of approximately 30° and 40° to the vertical (Fig. 6*c*). The patterns of the *thigh* are similar (Fig. 6*a* and *b*), but the lateral thigh surface has unit dimensions approximately 50% higher than the medial surface.

Lateral and medial surfaces of the *arm* also show differences in unit primary crease dimensions and the patterns near the joints are dominated by the creasing required for joint movement. At the *wrist*, circumferential creases

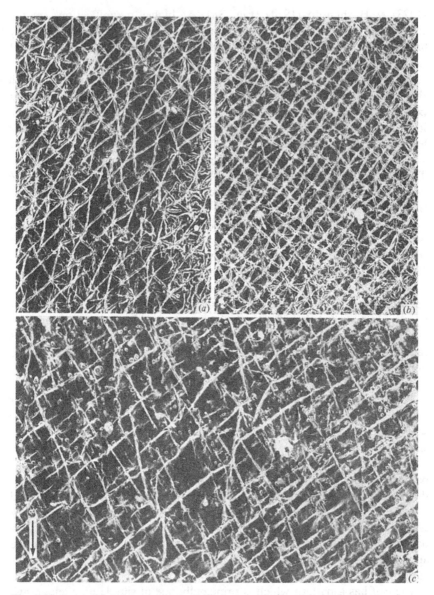

Fig. 6. Photographic enlargements (× 7.5) of crease patterns in stripped epidermal skin. (*a*) Thigh, lateral aspect, (*b*) thigh, medial aspect, (*c*) calf. The arrow indicates the direction of the long axis of the body. Note the very great differences in the crease pattern sizes on the leg.

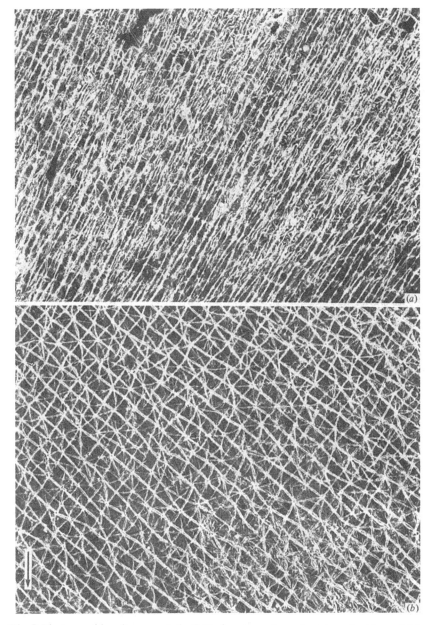

Fig. 7. Photographic enlargements (× 7.5) of crease patterns in stripped epidermal skin. (a) Forearm, lateral aspect, relaxed. Site approximately 4 cm distal to elbow joint. (b) Forearm, lateral aspect, with skin extended by flexing joint through 90°. Same site as in (a). The arrow indicates the direction of the long axis of the body.

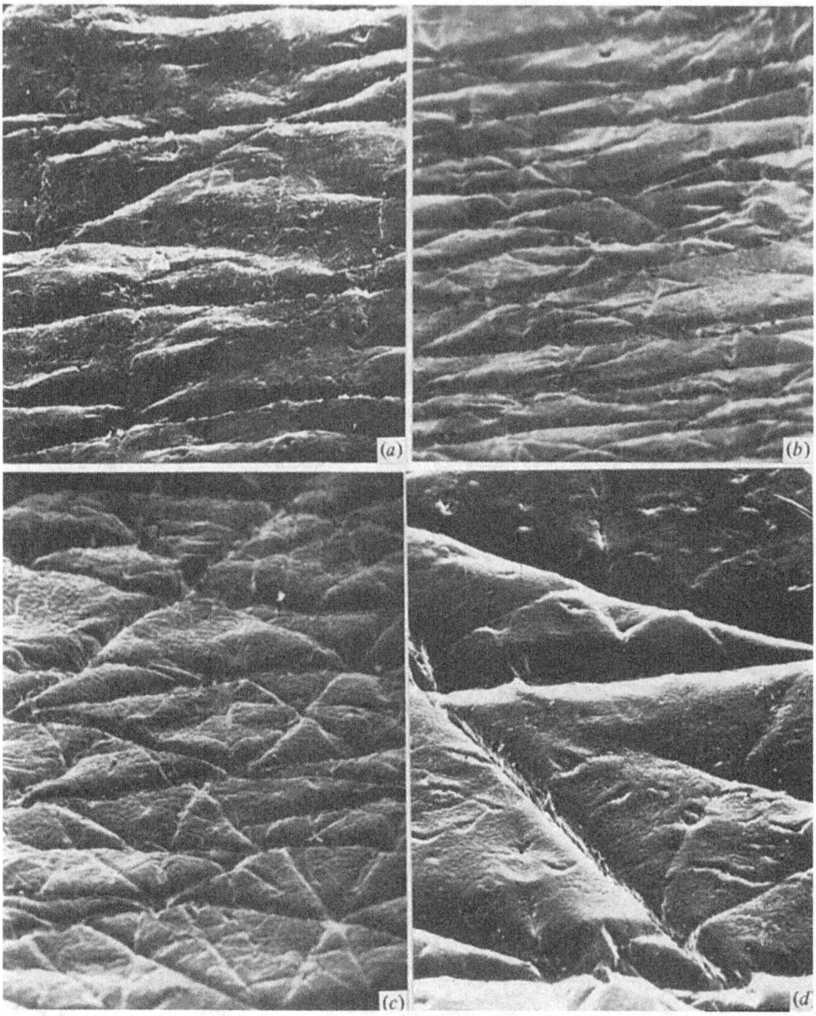

Fig. 8. Scanning electron micrographs of skin surface. (*a*) Forehead, (*b*) forearm, (*c*) shoulder, (*d*) back of hand.

predominate, but at the *elbow* the creases are inclined at about 60° to the long axis of the arm (Fig. 7*a* and *b*). The *back of the hand* has a surface pattern similar to that of forehead, although dimensionally different; the principal crease lines run circumferentially except at the margins. Lines normal to the major crease lines are not as evident on the back of the hand as elsewhere.

Although it is unlikely that there is a direct causal relationship between

Langer's lines (see p. 90) and all the crease lines of the stratum corneum, there are many coincidences in the directions of the primary creases and those of Langer's lines. Where it is possible to determine a 'principal' primary crease direction it is approximately parallel with the Langer line.

Scanning electron microscope images of replicas confirm the data obtained by cyanoacrylate stripped samples. Fig. 8 (a)–(d) illustrates areas similar to those above. Usually, scanning electron microscope images are obtained from tilted specimens, which gives a distorted view in the one direction but does emphasise the three-dimensional characteristics of the object. Even when photographed as a single image, it is possible to see both the actual (replica) form of the surface and by rotating through 180° to see the normal form of the skin surface. All illustrations given in Fig. 8 have been rotated through 180° to give a 'true' perception of the normal skin surface.

Scanning electron microscopy of fixed specimens has added much detail to our knowledge of the surface of skin. The cornified cells are characterised by a wrinkled surface with occasional fine nodular appurtenances (Orfanos, Christenhusz & Mahrle, 1969; Mishima, 1971; Dawber, Marks & Swift, 1972; Hashimoto & Kanzaki, 1975; Dempsey, 1979). Nodular cells have been found to be more common in disorders of the skin: psoriasis, icthyosis vulgaris, and some bullous pathologies. Mishima (1971) described the nodules on psoriatic squames as villus-like, approximately 400 μm in diameter with a spacing of 200–400 μm.

There are regional differences not only in the crease patterns of the skin but also in the size of the cornified cells. Individual cells from the surface of skin can be removed in a variety of ways. Wolf (1939) removed surface cells on adhesive-coated transparent tape. Hunter, Pinkus & Steele (1956) elaborated the method by dissolving the tape in a suitable solvent to provide a suspension of cornified cells. They measured the diameters of these cells and found that they ranged from 25 μm to 35 μm.

Plewig & Marples (1970) used a non-ionic detergent method, based on that previously adopted by Williamson & Kligman (1965), to obtain a suspension of corneocytes from the surface of skin. They used 1 ml of 0.1 % Triton X-100 in buffered phosphate in a cup placed on the skin to define an area of 3.8 cm². By scrubbing the skin surface with the blunted end of a plastic rod for 10–60 seconds, they obtained suitable samples. After air-drying a drop of suspension placed on a microscope slide, the cells were covered with immersion oil and inspected at high magnification by light microscopy. Measurements taken of two diameters at right angles to each other were used, assuming an elliptical shape, to calculate the area of individual cells. The choice of an ellipse as the optimum shape was adopted after considering various profiles, including that of a hexagon.

Cells on the *forehead* have been found to have the smallest area (746 μm²), whereas those of the *axilla* were the largest with an area of 1222 μm². The *abdomen, arm* and *thigh*, all areas of the body used to obtain grafts for surgical transfer, were found to have cells of a similar surface area, approximately 1100 μm². Cells on the *hand, forearm* and *scalp* were all smaller with values ranging from 800 to 875 μm².

Because the surface of skin is not smooth and the roughness appears to increase in disease, attempts have been made to measure the degree of roughness present. Nicholls, King & Marks (1978) used a replica technique to quantify skin surface irregularities. From a replica taken with Silflo impression material, they made a 'positive' by covering with DPX (R. A. Lamb, London) histological mounting medium. The surface contours were measured with a Surfometer (Planner Products Ltd) which could record micrometre changes in contour. Marks & Pearse (1975) had previously used this device on the 'internal' or fractured layer taken with cyanoacrylate, in an attempt to investigate adhesion between cells in the stratum corneum. Ishida *et al.* (1978) introduced more sophistication into measuring skin roughness, by using computerised surface tracing. Following the study of Hoppe (1979), Makki, Barbenel & Agache (1979) presented data reflecting the variability of skin surface contours. Makki (1980) found that the various amplitude measurements were not particularly sensitive, but spacing-dependent sets of measurements were useful for detecting angular variations of roughness.

Earlier Japanese studies had attempted to find quantitative data with which to express the rather vague concept of smoothness, a parameter of interest to cosmetic chemists. By contrast Marks and his colleagues had been interested in parameters which could be used in the assessment of dermatological treatment of skin, but at present it would appear that the methods so far developed are too onerous and ill-defined to be used for this purpose.

DERMATOGLYPHICS

The importance of the ridges on the fingers, hands, feet and toes was recognised long before they became significant in forensic work. Permanence of the ridge patterns throughout life was demonstrated by Sir Francis Galton in 1892 and it would appear that Sir William Herschel made a number of contributions towards establishing the scientific basis of the patterns as well as their unique characteristics on each individual. The word dermatoglyphics, which means skin carving, was first used by Harold Cummins in 1926.

The skin ridges are a normal appearance of the volar skin in man, apes and monkeys. They act as an antislip device and are thought to improve the sense of touch. Flexion creases are *not* components of dermatoglyphics. Only in rare instances are individuals born with smooth fingertips (Baird, 1964). The affected persons described all showed transient congenital milia which disappeared at about six months of age.

The ridge pattern is established in early foetal life and certain disturbances at about this time, whether due to heredity or environmental factors, are

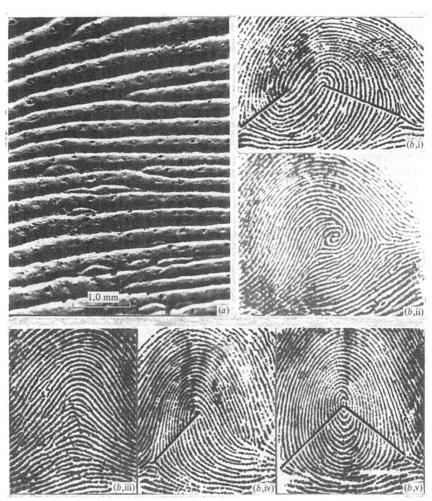

Fig. 9. (*a*) Scanning electron micrograph of fingertip ridges. (*b*) Fingertip patterns (enlarged) from ink impressions. Ridge counting is done along the straight lines connecting the core to the tri-radius. (*b*, i) Whorl, double loop; (*b*, ii) whorl, spiral; (*b*, iii) simple arch; (*b*, iv) radial loop; (*b*, v) whorl, symmetrical.

faithfully recorded as modifications in the ridged configurations. Skeletal malformations of hands and feet as in polydactyly, syndactyly and brachydactyly are all associated with such disturbances in or malformation of the ridge patterns, although Down's syndrome, characterised by a general retardation in growth affecting most parts of the body, is the disorder most commonly connected with these traits.

Each epidermal ridge has a single row of pores spaced at fairly regular intervals along the summit. Individual ridges frequently show discontinuities, branching or irregularities of direction (Fig. 9a). By adding grease, wax, ink or some similar substance to the surface of the fingers, a print may be obtained on any suitable flat surface. The usual method for taking prints is to use a non-staining greasy ink to produce clear black impressions; this inky substance can now be obtained on rolls of paper or plastic sheet for rapid finger-printing.

Patterns of finger prints are classified into three main types: arches, loops and whorls (Fig. 9b). *Arches* may be simple or tented. *Loops* are said to be ulnar or radial according to the direction they face. *Whorls* are of three types: symmetrical, spiral or double-loop. A *tri-radius* is the meeting place of three systems of ridges whose elements lie approximately parallel to one another, and the classification of patterns is determined by the number of tri-radii present. There is no tri-radius in a simple arch, there is one in a loop and two or more in a whorl. The *ridge count* is determined by drawing a straight line from the core or centre of a finger print pattern to its tri-radius and counting the number of ridges transected. It follows then that an arch which has a centrally situated tri-radius has a ridge count of zero.

In any individual, patterns vary from digit to digit and it is unlikely in a small group of unrelated people that the same sequence of patterns on the eight fingers and two thumbs will correspond. Loops are the commonest pattern and represent about 70% of all finger patterns in Britain. Whorls are found in 25% of the population, but arches in only 5%. Whorls have a maximum frequency on the thumb (digit 1) and the third (ring) finger (digit 4). The index finger (digit 2) most commonly has radial loops or arches while the little finger (digit 5) has the highest frequency of ulnar loops.

Differences in distribution of patterns on fingertips does occur in different races, but not exclusively in any particular ethnic group. The presence of ten ulnar loops occurs in about 35% of patients with Down's syndrome, but in only about 8% of the normal population (Thompson & Bandler, 1973). The total finger ridge count for all ten digits is an inherited metrical character. The accuracy of this statement has been demonstrated by correlation coefficients between counts and heredity characteristics, where in a random population sample the coefficient is 0.5 and in zygotic twins is 0.95.

On examining the palm through a low-power lens or by palm prints, it is evident that surface ridges course in different directions in different areas of the palm. Tri-radii can again be found. There is typically a tri-radius at the base of each digit except the thumb, and these have been designated *a*, *b*, *c* and *d* on digits 2, 3, 4 and 5 respectively. Another tri-radius can be found near the base of the fourth metacarpal (centrally at the base of the hand), although the exact position of this tri-radius is determined genetically; it is known as the axial tri-radius and designated *t*.

Suitable measurements from palm prints are governed largely by the elements normally found in forensic work on flat surfaces and on gripping prints. The palm laid on a flat surface will frequently leave evidence of the

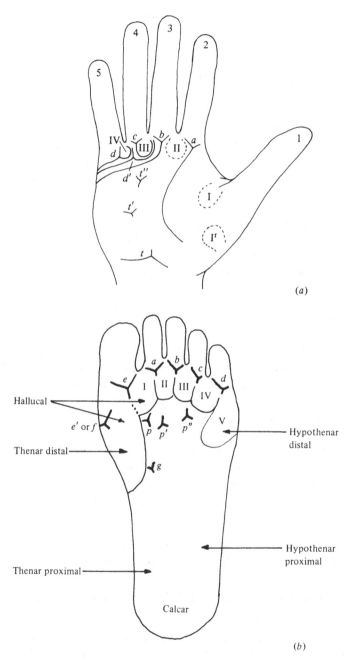

Fig. 10. (*a*) The arrangement of the tri-radii and some of the loops on the hand. The digits are given conventional numbering. (*b*) Configurational areas and tri-radii on the sole of foot. Notation is based on that of Penrose & Loesch (1969).

a, *d* and *t* tri-radii and so the angle between these three points is a useful measure. Similarly, where only part-prints are found, *a* and *b* are the most common tri-radii and the ridge count is again a useful measure. This measure is not so highly controlled by genetic factors, and early pre-natal environmental factors have some effect (Fang, 1950).

The epidermal ridge patterns on toes are similar to those found on fingers, but the frequencies with which they occur are different. Generally, arches are more numerous and whorls fewer. The traditional classification of patterns on the soles (as summarised by Cummins & Midlo, 1943) has been found to be inadequate for the purposes of genetic correlation. Consequently, Penrose & Loesch (1969) proposed a new method. Some of the topologically significant patterns are illustrated diagrammatically in Fig. 10 for both hand and sole. The frequencies of palmar pattern elements and also those of the sole are significantly different in men and women. In terms of dominance–recessivity it has been found that the loops I + I^r, II and IV and tri-radius on palm and I and V on sole relate to dominant characteristics, but the palmar loop II is the only recessive character (Loesch, 1974).

There is much detail to be found concerning ridge patterns and the reader should consult some of the literature available. The following are suggested: Cummins & Midlo, (1943), Schaumann & Alter (1976) or for a short review: Verbov (1970).

The production of good-quality prints either directly, photographically or in replica, is an essential prerequisite for all dermatoglyphic study (see Schaumann & Alter (1976), chapter 2, for a recent review). Routine prints onto paper can be made using a non-staining ink, but photographic records can be more useful, though contrast is a problem since skin itself is translucent. Callender (1974) proposed the use of carbon cream to emphasise the ridge patterns. The cream is rubbed into the skin surface with cotton wool and attempts to remove it are then made with a clean cotton wool swab. The process is repeated until the desired contrast is effected and the skin then photographed. However, one of the best methods for permanent record has proved to be Silcoset replication. The replica can be studied directly, photographed or inspected by scanning electron microscopy. Contrast enhancement, through the use of suitable creams or optically with directional light, is usually needed for photography of the white Silcoset.

BLASCHKO LINES AND OTHER SURFACE MARKINGS

BLASCHKO LINES

Blaschko lines are the pattern assumed by many different naevoid and acquired skin diseases. They were described and drawn by Blaschko in 1901 (see Jackson, 1976). These lines are to be distinguished from other surface patterns and do not follow the underlying nervous, vascular or lymphatic structures in the skin. Blaschko gave 83 examples of naevoid linear skin

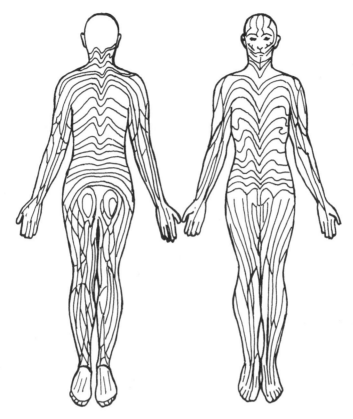

Fig. 11. Blaschko lines (redrawn): a system of lines on the body surface which linear naevi and dermatoses follow.

diseases and 63 examples of acquired linear skin diseases, which included examples of linear psoriasis, eczema, lichen simplex chronicus, lichen planus and scleroderma. In general, the naevoid conditions affect only a part of the body, but are characterised by being life-long, developing during intra-uterine growth or shortly after birth. Some of the acquired skin lesions have relatively short duration.

Blaschko put forward the concept that the lines always follow a set pattern or a system of lines on the surface of the human body. The numbers of lines shown in Fig. 11 are not precise, because in the thoracico-abdominal area the variations found do not allow an accurate assessment. In reviewing the original paper, Jackson (1976) contrasted other line patterns such as Voigt lines and Meirowsky lesion patterns, but eventually suggested that perhaps a form of human 'mosaicism' (based upon the ideas of Whimster, 1965) was the underlying cause of Blaschko lines, i.e. certain specific cells or groups of cells react differently from other cells due to some chromosomal abnormality.

STRIAE

Striae, or 'stretch marks', on the surface of skin have been observed in association with many debilitating conditions: tuberculosis, typhoid fever, rheumatic fever, and some chronic infections. The production of striae as a side effect of therapeutic administration of adrenal cortical hormones has led to support for a causal relationship (Poidevin, 1959; Chernovsky & Knox, 1964; Meara, 1964; Geschwandtner, 1973). Nevertheless, numerous sportsmen, particularly weightlifters, do develop striae coincident with rapid increase in muscle mass. The course of development is simple: initially lesions are flat, faint pink in colour and sometimes itch; with time the striae widen and lengthen and take on a vivid purple or red colour; after several years they fade, turning white eventually, and become depressed. The appearance and location are characteristic. They are commonly known to occur during pregnancy, but the aetiology of these striae gravidarum is not simple. Abdominal stretching during the later period of pregnancy is frequently taken to be the primary cause, but changes in steroid hormone levels could also be implicated. Since only a proportion of pregnant women develop striae the cause is more likely to be a combination of a number of contributing factors.

Analysis of striae supports the supposition that their origin is dependent on extensive changes in skin, and histological and ultrastructural studies show that collagen bundles within the dermis are or have been highly stressed in the striae regions (Arem & Kischer, 1980). Striae may be seen as a form of stress remodelling or growth of skin which includes dermal scarring perhaps under the influence of steroids.

MONGOLIAN SPOTS AND VOIGT LINES

Many infants of Mongoloid or Negroid extraction have slatey blue macular areas of pigmentation known as Mongolian spots. They may also have sharply defined lines between the dorsal and the paler ventral surface of the arm, sometimes called Voigt lines. Mongolian spots are occasionally found in young Caucasian infants.

The spots vary in diameter from 1 to 15 cm or more, and are commonest on the sacral region and buttocks. They are often present on the back and shoulders, and sometimes on the arms and legs, but are rarely found on the face. Larger spots may be formed by a coalescence, so that large areas of the back may be blue.

The blue colour of the skin is due to the particular optical relationship between collagen and melanocytes. The shorter-wavelength blue light is reflected by the surface-parallel bundles of collagen in the subjacent dermal region while the longer green and red wavelengths are absorbed by the melanin (Findlay, 1970). The origin of the spots lies in the failure of the melanocytes to migrate from the neural crest through the epidermal–dermal junction during embryonic development. The completion of this process in early childhood causes the spots to disappear. In a study of West Indian children, Vollum (1971) demonstrated that the spots were present in 95% of children under 1 year of age, but the percentage decreased linearly with age so that only 3% of children aged 10 retained their spots.

Matsumoto (1913) first used Voigt's name to identify the lines which mark the differences in pigmentation between the darker extensor and the paler flexor surface of the arms. These correspond with the anterior axial lines which mark the junctions of the dermatoses or areas of skin supplied by individual spinal nerves. They extend from the position of the second costal cartilage on the sternum, laterally over the pectoralis major to the anterior border of the arm and then medially across the flexor surface of the forearm to the mid-point of the wrist. Minor differences in the line patterns have been found, particularly along the anterior border of the arm.

COLOUR

Three factors influence the colour of skin: (1) the state of the circulation, (2) the pigments present, and (3) the thickness of the epidermis. Pigmentation derives not just from the presence of melanocytes in the basal layer of the epidermis, but also from migration into and around the epidermal cells of the dense melanocyte granules, the melanosomes. The many variations in skin colour show a regularity in world-wide geographical distribution and appear to be related to four quadrants of longitude (Roberts, 1977). These take on an adaptive role against ultraviolet radiation. Variations occur even among the Caucasian group, which may be explained in terms of different genotypes of cell, giving rise to melanocytes producing morphologically different melanosomes. Ultraviolet light stimulation may also cause melanocytes in one person to produce more melanosomes than in another person of the same region and race. Finally, a form of mosaicism may also exist in which melanocytes are grouped together focally to cause patchy coloration.

Different melanosomes have been encountered in skin showing distinctive features. The inter-freckle dermis of red-haired people contains melanocytes which have granular organelles containing particles similar to those known as pheomelanosomes (Breathnach, 1964). Pheomelanosomes are membrane-limited intracellular bodies with microvesicles and filaments. The differing density of the microvesicles gives the body its characteristic appearance: the normal melanosome is usually seen in the electron microscope as a uniformly dense body within which the individual particles are difficult to distinguish.

Despite the continuing interest in melanocytes over the past 40 years, most of the current knowledge rests on studies of animal skin. It would appear that in mice the entire melanocyte population comes from only 34 initiator cells (Silvers, 1979). The commitment to form melanocytes occurs within the neural crest before migration; the genes responsible for differentiation become active very early during gestation. The cellular environment is also important and may control melanocyte performance throughout its entire life cycle.

The changes in skin which influence the function of epidermal melanocytes are evident in some people with piebaldism (Quevedo & Fleischmann, 1980). Melanocytes are absent from the epidermis and hair follicles of the characteristic white forelock, but are present in the hypermelanotic areas within the hypopigmented patches of the skin of the trunk and extremities (Jimbow *et al.*, 1975). These melanocytes produce abnormal melanosomes. The available evidence suggests that the melanoblasts have migrated from the neural crest to all regions of the skin, but have failed either to survive or to differentiate.

Normally melanocytes disappear from the epidermis with increasing age. The epidermis of non-exposed buttock, abdominal and medial arm skin of Caucasians undergoes a substantial reduction in numbers of melanocytes throughout adult life. In the chronically sun-exposed lateral aspects of the arms, the number of melanocytes is substantially higher than in the medial aspect for each age. However, it is uncertain at the present time whether the age-dependent decline in melanocytes represents a true loss of cells or simply a conversion to an amelanotic form.

At all ages, the exposure to ultraviolet light whether natural or from 'sun lamps' results in a darkening of the skin. New Guinea natives, born with relatively little melanin, rapidly acquire pigment in the first 6 months of life, so that the forehead, arm and axilla are as dark as in the adult (Walsh, 1964). The response to stimulation is of two types: the first, generally called primary melanisation, consists of an erythemal response (e.g. sunburn) followed by the formation of a new pigment (melanogenesis) and the migration of the melanin granules through the epidermis; the second, known as immediate pigment darkening, is evoked by longer light wavelengths. The darkening of skin pigment was first reported by Hauser (1938) and in more recent studies (Pathak, Riley & Fitzpatrick, 1963) the phenomenon of darkening was explained in terms of the absorption of light energy by melanoprotein molecules, and the transmission of this energy by conduction bands. Fading and lightening of the colour of immediately darkened melanin would appear to be due to the redox system of the cell.

Melanisation of the skin is a response that can be induced by a variety of stimuli: injury caused by ultraviolet light, by X-rays, by thorium radiation, by the photodynamic and photosensitizing dyes and the furocoumarins. Active proliferation always accompanies melanisation and is manifested by the rapid increase in mitotic activity of the basal cells of the epidermis. Many factors which affect cell turnover rates in skin also affect skin colour. Genetic factors are perhaps the most obvious, white and black being determined by inheritance. The categories of 'black' and 'white' skin, however, may relate to both cultural and physiological differences. Blood flow through skin, particularly that near the epidermal–dermal junction, influences skin colour. Indeed, Harburg *et al.* (1978*a*, *b*) suggest that blood pressure (a socially related stress factor) is associated directly with skin colour and may have important social consequences. In order to exclude the effects of blood flow (as distinct from pigmentation) on the erythemal response, Henschke & Schulze (1939) pressed a glass slide onto the skin surface in order to collapse the dilated skin capillaries. Any attempt to measure skin colour must take into account the erythemal response of the tissue.

Various trace elements also affect the degree of melanisation of skin and hence its colour. The loss of pigment found in copper deficiency states in sheep has been attributed to a breakdown in the earlier stages of melanin synthesis, which are catalysed by the copper-dependent enzyme tyrosinase (Underwood, 1971). Very high zinc levels are found in heavily melanised structures like the choroid of the eye and pigmented moles. Hyperpigmentation of skin lesions has been reported in patients with abnormalities of metabolism, such as Wilson's disease and haemochromatosis. High serum copper levels are found in a number of hyperpigmentary conditions of the skin such as xeroderma pigmentosum, but melanosis accompanying pregnancy or intake of oral contraceptives would indicate that trace elements are not the only factor. Similarly, colour changes in hair found in children with protein deficiency were not related to copper nutrition (Gopalam, Reddy & Mohan, 1963).

Metal ions can produce skin discolouration in a number of ways. The deliberate introduction of heavy metals into the skin takes place in tattooing, where salts of cadmium, chromium, cobalt and mercury are frequently used (see Chapter 6). In a variety of industrial situations, metal particles can become embedded in the skin, but in the majority of cases the introduction of the metal does not induce melanosis. It has been with the excessive use of medicaments containing arsenic, bismuth, gold or silver that melanosis has long been known to be associated. The increase in melanin is noted particularly in the basal cell layer and is thought to be caused by the binding of these heavy metals to tissue sulphydryl groups (Lorincz, 1954). Hyperpigmentation resulting from the use of mercury in cosmetic creams is probably quite common (Burge & Winklemann, 1970). Initially there is a bleaching of the skin, probably caused by inhibition of melanin formation, but over a long period of time an increase in skin colour is usually observed. This increase could again be due to the binding of mercury to the sulphydryl groups.

TEMPERATURE

Measurement of skin temperature reveals significant differences between one part of the body surface and another (Table 1). Undoubtedly, control of circulation is a major factor. That the response is adaptive has been known for a long time. Indeed, it is generally accepted that people chronically exposed to cold exhibit peripheral heat-regulating responses which may enable them to survive in their environment. Not everyone, however, can adapt, as has been shown in many laboratory controlled experiments such

TABLE 1. *Skin temperatures: nude subject after 20 minutes at room
temperature (23 °C); rectal temperature 37 °C*

Body area	Skin temperature (°C)
Forehead	33.4
Clavicle	33.6
Right thorax	32.8
Mid-abdomen	34.2
Arm	32.8
Palm	32.8
Lumbar area	33.3
Knee	32.5
Calf	33.2
Sole of foot	30.2
Toe	31.0

as those conducted by Miller & Irving (1962) where the hands of subjects were cooled. But even when given the opportunity to adapt, people from a temperate climate do not function as well as Eskimos in a very cold climate.

Body surface temperature may change as a result of response to an internal stimulus or change of overall condition of the body. A change in the total blood volume may be reflected in the amount of blood in the skin vessels. After haemorrhage there is a decreased flow of blood through the skin, a reduction in the total amount of blood in the skin and a feeling of coldness. In some instances of strong sensory stimulation, some emotional states and the initial stages of muscular exercise, redistribution of the blood in various parts of the body may induce similar changes. In other emotional states, blushing can occur with an increased blood flow in certain areas such as the face and neck.

Local transient changes in blood flow and hence surface temperature, can be readily induced. Mild stroking with a blunt point results in a local contraction of the capillaries and an emptying of the superficial venous plexus, the white reaction. Firmer stroking results in Lewis's triple response: (a) a red line with transient white edges, due to central capillary dilatation edged by capillary constriction; (b) a spreading red flare due to arteriolar dilatation; (c) the formation of a weal due to the increased permeability of the capillaries.

The advent of infra-red colour thermography has made it possible to visualise the surface temperature of the human body over large areas of skin and thus observe directly the effects of movement, varying air flow and changes in skin blood flow on surface temperature (Fig. 12). At ambient temperatures between 10 and 20 °C the range and distribution of skin temperatures are more variable (± 10 deg C) than they are in a warmer

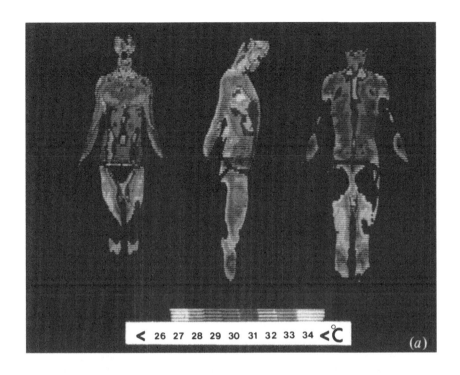

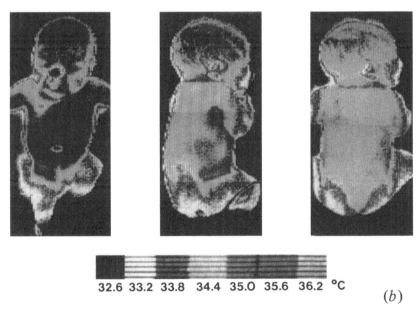

Fig. 12. (a) Infra-red thermograms of temperature distribution on an adult male in a climatic chamber at 11 °C, showing cold knees, buttocks, backs of arms and warm spine. (b) Infra-red thermograms of temperature distribution on a young child at an environmental temperature of 32 °C. Typically, children have a warm trunk and head and are cooler towards the extremities. (Illustrations kindly provided by Dr R. P. Clark. From Clark & Stothers (1980) and Clark, Mullan & Pugh (1977).)

environment. At about 32 °C, a thermally neutral condition, both adults and infants have an overall temperature range of only 4 deg C. At rest, the conductance and metabolism of structures underlying the skin have minimal effect on the temperature patterns recorded (Clark & Stothers, 1980).

There is a redistribution of skin temperature during exercise, when skin temperatures are found to be coincident with the surface outlines of active muscles (Clark, Mullan & Pugh, 1977). Removing sweat does not change the colour patterns observed by infra-red thermography and indicates that the presence of surface water is not highly significant. This is exemplified in the case of the forehead, which is notably cooler during running but sweats more profusely than warmer areas. Under resting conditions, higher temperatures are found over the spine and correspond with fat deposition patterns, although elsewhere temperatures are generally lower over body prominences and fat depots. The elevated temperatures over the spine relate to deposits of brown or yellow fat.

For the measurement of mean body temperature, the selection of site is important. The surface of the upper arm and, to a lesser extent, the upper thigh, give values which represent the mean value most closely. The upper abdomen and head are generally warmer and less responsive to environmental change.

SOFTNESS

Skin smoothness and softness are two parameters commonly accepted as indicators of a good healthy skin. As skin dries out, so it becomes more flaky and relatively rough and hard, and many procedures have been tried for evaluating the efficiency of treatments to soften and smooth skin. Those that have applied judgemental approaches have used rating scales (e.g. Imokawa, Sumura & Katsumi, 1975a, b). Others have attempted to measure some physical property directly on the skin. Prall (1973) reviewed the instrumentation available in terms of the 'smoothness' concept and developed a simple device for measuring scratch hardness.

The perception of skin smoothness and the physical factors associated with smoothness are complex. The neuro-physiological processes of touch have received considerable attention, with aspects such as adaptation, discrimination, velocity of movement, greasiness and kinesthesis all playing a role. Edge and contrast effects are in general felt more intensely than steady or frequent stimulus. Thus, the stick–slip situation will be more noticeable than a steady movement across the surface, but the contribution of friction has only been investigated in superficial terms, since the classical laws of friction cannot wholly apply to skin.

The term 'softness' is also applied uncritically to skin, often as a synonym for 'smoothness', and is related to the term 'scaliness'. The rate of desquamation from the surface of skin affects the surface in a number of ways: loss

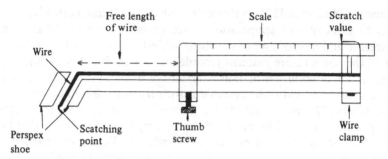

Fig. 13. Instrument for measuring the scratch hardness of skin. (Re-drawn by permission of Dr J. K. Prall.)

from a small local area leads to a consequent increase in 'roughness', but at the same time reduces the number of cornified cells overlying the softer cells of the epidermis; by contrast, a decreased rate of loss may lead to a 'harder' surface, and also to a drier, flaky, or 'rougher' surface. Difficulties in evaluating 'scaliness' or 'hardness' arise because they involve a variety of parameters including dryness, roughness, horniness, flakiness and friction. Some simple measures can be taken, however, and Prall's device for measuring the scratch hardness of skin is one example (Fig. 13). The tip, of 5×10^{-4} inches radius, is drawn across the skin surface repeatedly at increasing load until a scratch just appears. The results, however, are still subjective since identification of the scratch depends upon the adeptness of the observer in recognising the change in light pattern scattered from the surface.

Indentation is a better way to measure 'softness' and suitable techniques are well established. Indeed, most people test for softness by prodding with the tips of their fingers or with a pencil, etc. The criterion used is the depth to which indentation occurs. Weinstein (1978) used a device comprising a piston within a cylinder whose vertical movements were detectable to 10^{-4} inches. With this equipment he was able to show that soaking skin in water or lotion had an immediate effect in enhancing skin softness, but after a period of 5 minutes the effects of the water had largely disappeared. Since skin indentation is a measure of the total soft tissue response, including that of the deeper tissues, results from this type of experiment have limited scope for interpretation.

Weinstein also devised a method for measuring skin smoothness *in vivo* using a sled of deglazed thermoplastic drawn by a constant-speed motor (5 mm/min). The sled was loaded by means of a rod with a small ball-bearing point resting on the centre of the sled. With the skin and sled maintained horizontally, direct measurement of starting friction and constant velocity friction could be made. He found that water and very dilute detergents produced immediate reductions in skin friction whereas cosmetic lotion, such as shaving lotion base, did not. In further studies he found that dermabrasion of dry skin produced a smoother surface, but not after treatment with water or lotion. Apparently, water or lotion had had some effect on the skin which was not simply that of a lubricant.

Packman & Gans (1978) used a consumer test involving large numbers of independent observers to demonstrate the efficiency of skin moisturisers. Three out of five moisturisers tested were judged better than water, but water itself proved to be 'quite good' and better than no treatment at all. The role of water in determining the quality of skin at the body surface is one of primary significance.

A level of between 10 and 20% water is needed to maintain skin in a soft pliable condition. The normal supple appearance is due largely to the ability of the stratum corneum to bind water. In conditions of low relative humidity, the amount of water diffusing through the stratum corneum is insufficient to replace the water lost. On the other hand, hydration of the skin rapidly increases its permeability to water as well as affecting its mechanical properties. Dry stratum corneum presents high resistance to penetration by water.

The total transepidermal water loss in 'comfortable conditions' for the normal adult varies between 85 and 170 ml/day, added to which sweating contributes about 300–500 ml/day. A large proportion of the water is lost from the hands and feet.

Most studies have attempted to measure evaporatory water loss per unit area of total skin surface, although in very young infants respiratory water loss is frequently included. The data available tend to show that water loss through the skin of infants is between 33 and 50% lower than that found in adults, but in low-birthweight children water loss may increase by about 120% over the usual level for infants.

Measurement of water loss is usually made using techniques for evaluating the water content of gas or air. The most common method employed is an electrolytic water vapour analyser (Gasselt & Vierhout, 1963; Spruit & Malten, 1966; Thiele & Senden, 1966; Grice, Sattar & Baker, 1972; Rajka & Thune, 1976). Water vapour can also be measured by thermal conductivity (Rosencrants, 1930; Adams, Funkhouser & Kendall, 1963; Cherry, 1965; Cohen, 1966; Spruit, 1967), by electrohygrometer (Bettley & Grice, 1965; Baker & Kligman, 1967; Lamke, 1970), by infra-red absorption (Palmes, 1948; Staats, Foskett & Jensen, 1965; Goodman & Wolf, 1969; Johnson & Shuster, 1969) and even by changes in dew point temperature (Brenglemann, McKeag & Rowell, 1975). For a review of these methods see Nilsson & Oberg (1979).

Wildnauer & Kennedy (1970) measured the transepidermal water loss (TWL) over a defined area of the body surface. They used equipment in which dry air was passed over the skin surface via a sampling chamber subtending 16 cm^2 area of skin (A). The relative humidity (r.h.) and temperature were monitored so that r.h. represented the water vapour picked up from the skin surface. Volumetric air flow (AF) was maintained at constant rate during each experiment. Calculation of TWL was made according to the formula:

$$TWL = \frac{r.h.}{100} \cdot D \cdot AF \cdot \frac{1}{A}$$

where D is the weight of water per litre of saturated steam at the temperature of the air passing over the skin, in mg/l.

The r.h. was not measured until the hygrometric system and the skin surface had equilibrated. Excessive emotional sweating was found to be transient. Similarly, thermoregulatory stimulation appeared as rapid and large increases which did not return to the baseline TWL.

Wildnaur & Kennedy found that the transepidermal water loss in adults was 0.27 mg/cm² per hour when measured on the upper back. In the newborn, they reported values of 0.18 mg/cm² per hour for the upper back and 0.17 mg/cm² per hour for the buttock, so confirming the lower rate of loss per unit area of skin in the newborn. Ambient temperature of the site was 70 °F (27 °C).

Hammerlund *et al.* (1977) measured water loss from the body surface of newborn infants at an ambient temperature of 34 °C. They found average water loss to be 8.1 g/m² per hour (0.81 mg/cm² per hour) but also reported that a change in humidity from 20 to 60% lowered the rate of water loss by 40%. From these and other reports it seems that relatively small changes in temperature and humidity can alter markedly the rate of water loss from the skin.

Hydration of skin increases the thickness of the epidermis several-fold and, upon immersion, swelling begins almost at once. Measurements of the increase in thickness can be made within a few minutes. The stratum corneum will continue to swell for up to 3 days, and may reveal an increase from 5% to 50% water. It can absorb five to six times its dry weight.

Despite the lower transepidermal water loss found in the newborn, foetal skin is much more permeable to water, but its permeability decreases with increasing age. Diseased skin is also more permeable to water than normal adult skin: e.g. in eczema, values are about 8 to 10 times greater. In such cases, loss of water from the whole body surface may approach 2 litres per day. Thus, apart from its effect on water and salt metabolism, the evaporation of this amount of water from skin requires over 1000 calories. Under certain conditions this could impose an additional stress on patients with diseases of the skin.

Some steroids, in particular oestrogens and pregnenolone, have been reported to produce beneficial effects on ageing skin by hydrating it (Sternberg, Levan & Wright, 1961; Grant, 1969). The site of action is, however, still a matter for discussion. From experiments in which corneal cells were powdered in liquid nitrogen, Middleton (1968) suggested that this site was the cell wall that contained the lipids associated with water binding. It now appears that in human and rabbit skin a hexane-soluble lipid is the major regulator of passive water binding. By contrast, burned skin contains 30% or less of the normal 'water-holding' lipid and transmits up to four times the water passed by intact skin. Hormones act principally on cell membranes and presumably assist the process by which water is held. Substances which come between lipids and other membrane molecules might be expected to reduce the water-binding capacity.

Soaps and detergents are perhaps the most damaging of all substances routinely applied to skin. Treatment of isolated human or animal callus with soap or detergent solutions can reduce the ability to hold water in humid atmospheres (Blank & Shappirio, 1955; Singer & Vinson, 1966). Dilute aqueous solutions of anionic surfactants were found to increase the rate of water loss through human epidermis (Bettley & Donaghue, 1960; Bettley, 1961, 1963, 1965; Bettley & Grice, 1967). Damage to skin appears to be a general property of surfactants (Middleton, 1969; Scheuplein & Ross, 1970), and the presence of anionic surfactants within skin greatly reduces the amount of 'bound water' as measured by desorption experiments (Scheuplein & Morgan, 1967). The major differences between the effects of different anionic surfactants on 'water binding' may be explained in terms of different degrees of interaction between lipids and water-binding substances, since neither pH nor surface tension has a direct relationship to permeability. Indeed, changes in pH of 1–10 have little effect on the swelling and hydration characteristics.

Using dry electrodes to measure the resistance and capacitance to high-frequency current of 3 MHz, Tagami *et al.* (1980) demonstrated that scaly lesions of the skin had a decreased water-holding capacity despite increased transepidermal water loss. It would appear, therefore, that the water-binding capacity of skin is not directly related to the rate of transepidermal water loss, to pH or to surface tension, although each of these may be important.

The precise site of the 'water barrier' in stratum corneum has been the subject of much conjecture. Earliest suggestions placed it between the stratum germinativum and the stratum corneum (Rothman, 1954; Stoughton & Rothman, 1959; Matoltsy, 1967), but it was also placed in the lower portion of the corneum (Szakall, 1951; Blank, 1953; Monash & Blank, 1958). More recent studies indicate that the barrier to penetration is the entire stratum corneum (Kligman, 1964; Matoltsy, Downes & Sweeney, 1968; Parmley & Seeds, 1970; Scheuplein & Blank, 1971).

After damage, 'water barrier' renewal is a complex process. If the stratum corneum is removed, a temporary barrier is set up by rapid conversion of granular cells into parakeratotic cells. As the parakeratotic layer thickens, the water loss gradually decreases. The temporary barrier persists until the regenerating epidermis forms normally keratinised cells. Within 3 days of injury, water vapour loss from the skin surface is reduced to approximately normal. Total regeneration of the epidermis takes about 2 weeks (Matoltsy, Schragger & Matoltsy, 1962; Spruit & Malten, 1966).

The importance of water in relation to flexibility was first noted by Blank (1952). The dependence of the elastic properties of the corneum on relative humidity was demonstrated by Middleton (1968). Later, using isolated stratum corneum, Wildnauer, Bothwell & Douglass (1971) showed that the elongation to fracture increased with increasing relative humidity, whilst the load at fracture was reduced. The dependence of water content and elastic properties on humidity, suggested that skin chapping could be related to humidity. After consideration of temperature changes both above and below ambient, Middleton & Allen (1973) found that extensibility was lower at reduced temperatures although the corneal water did not decrease. They concluded that skin chapping is the result of a combination of low corneum temperature and water content and that greater flaking and chapping of the surface corneum are caused by a gradient of temperature or water content across the corneum in a cold or dry environment. More recent studies have confirmed (Van Duzee, 1978) that the elastic modulus is independent of temperature, but is a function of water content.

Idson (1973) suggested that it is more important to consider the thermodynamic activity of water in the barrier phase than the amount of water present. Gaul & Underwood (1952) had obtained values for diffusion rate of water in skin of 0.5×10^3 cm/h at 25 °C or a flux of 0.2 mg/cm² per hour. They found that transepidermal water loss by passive diffusion varied between 0.2 and 0.6 mg/cm² per hour. They also suggested that the variation was largely related to site: e.g. average values found for the abdomen were 0.35 mg/cm² per hour and for the thigh 0.6 mg/cm² per hour. Permeability also increases with hydration. Since then, many measurements have been recorded and summarised (Shahidullah, Raffee & Frain-Bell, 1968; Nilsson & Oberg, 1979), with values varying from 0.1 to 9.0 mg/cm² per hour (1.0 to 90 g/m² per hour) depending upon the method used and the site investigated. Maximum rates have been found on the hands (50–90 g/m² per hour), feet (28–48 g/m² per hour), head (18–35 g/m² per hour), and trunk and limbs (6–11 g/m² per hour), but these figures included loss from sweat glands (Lamke, Nilsson & Reithner, 1977a).

It remains the case, however, that regardless of the increase in measured permeability, the highly hydrated stratum corneum is relatively impermeable to water. It is an effective and stable barrier. The hydrated stratum corneum has a diffusional resistance some 10000 times greater than an equivalent of water (Scheuplein, 1966). The energy of activation for the diffusion of water through highly hydrated stratum corneum has been calculated to be 13–16 kcal/mole (Mali, 1956; Scheuplein, 1965). These calculations indicate that diffusion of water within tissue is orders of magnitude less than the auto-diffusion rate of water (Scheuplein & Ross, 1970). This can readily be demonstrated by measuring the resistance of normal forearm skin to water transport before (150–160 s/cm) and after removing the stratum corneum by Sellotape stripping (0.4–1.25 s/cm.).

By comparing drug diffusion coefficients in water and in hydrated stratum corneum, Scheuplein (1966) obtained data indicating that the diffusion coefficients through hydrated stratum corneum ranged from 10^{-9} to 10^{-13} cm/s whereas the rates in water were 10^{-5} to 10^{-6} cm/s.

2

THE APPENDAGES OF SKIN

NAILS

The finger and toe nails are modifications of the epidermis in which a very thick horny layer replaces the keratinised layer of cells (stratum corneum). The flat, horny plate is known as the *body of the nail*; it rests on the *nail bed* (Fig. 14*a*). On the proximal side, the nail grows out from a fold in the skin, the *proximal nail fold*. The edge of this fold is continued over the upper surface of the nail as the *eponychium*, while the free edge of the nail projects over the underlying skin, *hyponychium*, and the sides of the nail lie in lateral grooves.

The greater part of the nail body is translucent and pinkish in colour due to the blood in the nail bed. This translucent portion overlies a thin matrix of epithelial cells. The small, whitish, semi-lunar part, the *lunule*, partly covered by the nail fold, represents the region from which formation of nail takes place. This generally held and traditional view of nail formation from the proximal part of the floor below the lunule has been challenged. Lewis (1954) produced evidence, since supported by histochemical and electron microscope studies, that nail is formed in three layers which he called the dorsal, intermediate and ventral nails, (Fig. 14*b*). The dorsal nail was said to be formed from part of the overlying fold, the roof, and a small part of the underlying region, the floor, of the nail fold, but from abrasion and wear part of this portion of the nail may be lost before it reaches the free edge. The intermediate nail is formed from the remainder of the traditional matrix, while the ventral nail arises from the whole of the nail bed. An older theory, illustrated in Fig. 14(*c*), suggested that the nail bed is generally sterile except for a small portion close to the point of separation of the nail from its bed. This area was given the name *Solenhorn* (Sammon, 1972).

While there is support for all three theories, none of it is definitive. Nail ablation frequently results in the growth of nail spicules at the distal end of the nail bed and in certain pathological conditions nail appears to develop from all parts of the bed. In mammalian claw, formation is in two layers, so given appropriate stimulus, human nail could have the ability to form in more than one manner.

The colour of the nail in the matrix region was thought to be due to an inherent opacity, but avulsion of the nail plate reveals a lunule in the underlying tissue. It would appear that the nail substance itself is indeed opaque when first formed, but becomes translucent as it passes from the nail matrix to the nail bed. Added to this change, it would appear that the major

31

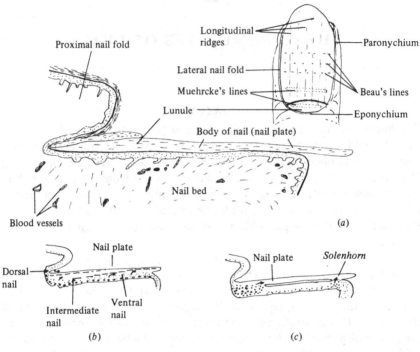

Fig. 14. Human nail. (*a*) Plan and longitudinal section view. (*b*) Lewis's hypothesis of nail formation in which three layers of nail are identified. The dorsal was thought to be worn away leaving only the intermediate and ventral layers as free nail. (*c*) Earlier hypothesis (Boas's theory) of nail formation in which the nail body is formed at the root of the nail fold and proximal nail bed. Secondary formation took place at the *Solenhorn*.

part of the nail matrix has a scanty capillary network and a loosely textured dermis with a few collagen fibres. The disappearance of the lunule has been associated with an increase in dermal collagen in the nail bed enhanced by an increase in the immediately subjacent capillary network which gives the pink colour. The eponychium and the proximal nail fold when not habitually pushed back for cosmetic reasons afford an excellent site for the examination of human capillaries.

The surfaces of finger nails are not smooth, but bear longitudinal ridges on both the upper and lower surfaces (Fig. 14*a*). These ridges are formed as the nails grow and correspond to parallel dermal ridges. The dermal ridges are modifications of the dermal papillae that underlie the epidermal ridges of the hands and feet. Thus, the longitudinal ridges of the nail correspond to the structures which produce finger-print identification (Thomas & Baert, 1967). Normally, nails are inspected in oblique reflected light to emphasise the relief patterns of the nail surface, but recently, Apolinar & Rowe (1980) developed a method to make use of the polarising properties of nail. Their method was to embed nail clippings in an acrylic resin and then thin them by grinding from the convex side. The concave side retains the better ridges, those on the outside being subject to more wear. After thinning and polishing, the nail was inspected in a comparison microscope with the capability for orthoscopic polarised light observations. The microscope was equipped with rotating stages so that maximum contrast between the bands of interference colours could be obtained. Apolinar & Rowe claim very accurate association of bands from the same individual to provide a certain identification; each finger had its own characteristic pattern.

Other surface markings on nails include Muehrcke's lines, which are paired narrow white bands associated with hypoalbuminaemia (Muehrcke, 1956). They are frequently reversible. Transverse bands are also associated with certain chemotherapeutic agents such as doxorubicin. Cyclophosphamide and bleomycin may also induce pigmented nail bands. Transverse ridging (Beau's lines) appears to be associated with abnormal keratinisation and in the sub-epidermal tissues there may be inflammatory exudate: lymphocytes, plasma cells and polymorphs.

Measurement of the rate of growth requires that specific zones are identified either by marking the nail or by means of natural grooves such as that which frequently appears just distal to the apex of the lunule (Fig. 14a). From cross-sectional studies of the population, it is relatively easy to determine age-related changes, but longitudinal studies are tedious and have only been reported for relatively short periods of time. The rate of linear nail growth increases until well into the third decade of life (Orentreich, Markofsky & Vogelman, 1979). From 25 years onwards the rate of decrease is approximately linear at about 0.5% per year. The rate ranges from 47 mm/year in the early twenties to about 29 mm/year by age 70.

Various factors affect nail growth rate. Skin temperature appears to be one of the more important transient factors. Rates have been measured in adults from 1 μm/h at 17 °C to 12 μm/h at 32 °C, but the relationship between growth rate and temperature is not linear. In one series of measurements, maximum rate appeared to occur between the temperatures of 24 and 28 °C, with rates of 3 μm/h to 9 μm/h respectively. Natural variations in rate also indicate an influence by diurnal rhythms and the presence of a long-period rhythm with a cycle of about seven years, but these have a relatively small effect. Other factors that make for a faster growth rate are: third digit, males, piano-playing, typing, handedness, premenses, pregnancy, psoriasis, onycholysis, hyperthyroidism, hyperpituitarism, arteriovenous shunt. A slower growth rate is associated with: first and fifth digits, females, ageing, smoking, lactation, acute infection, pneumonia, malnutrition, hypo-thyroidism, peripheral neuropathy, decreased circulation, congestive heart failure. Height and weight, skin colour, use of nail polish or varnish, dietary supplements, corticosteroids in moderate doses, minor illness or surgery have no effect.

Measurement of physical properties of nails has been carried out by Maloney, Paquette & Shansky (1977). Their data confirmed earlier reports (Baden, 1970) of flexion strength and have largely defined the parameters for future studies. Since all biological material has complex mechanical properties it is important always to give the conditions under which the particular property has been measured. Flexural strength of nail measured at a deflection of 0.5 mm with cross-head speed of 1.2 mm/min is generally found to be between 3.4×10^7 N/m² and 1.2×10^8 N/m². The tensile strength of nails measured at a cross-head speed of 5 mm/min is 3.1×10^7 N/m² to 1.2×10^8 N/m². Resistance to tearing measured at the slow cross-head speed of 1.2 mm/min varies between 1.9×10^6 N/m² and 4.7×10^6 N/m².

HAIR

Wherever hairs are found on the surface of skin, they are present as one part of a hair follicle unit (Fig. 15). This unit consists of the hair follicle, the hair shaft, one or more sebaceous glands and small, smooth-fibred bundles of the arrectores pilorum muscles, which upon contraction pull the hair into a more erect position.

The solid cords of cells which extend downwards into the mesenchymal layer at about the twelfth week of foetal life (Fig. 2c) quite soon show the bulbous swelling characteristic of the hair root and into this, from the lower surface, mesenchyme penetrates to form the papilla of the hair bulb. Several lobes of a single sebaceous gland or separate glands later develop as pockets from the upper part of the primitive follicle. This is followed by the development of a strand of smooth muscle differentiated from the mesenchyme.

The base of the hair follicle consists of undifferentiated, multi-potential cuboidal matrix cells. These give rise to four different cell types: those of the medulla, cortex, cuticle and inner root sheath. In the centre of the hair follicle the ascending cells develop small granules in their cytoplasm and become the medulla of the hair. Around the medulla, spindle-shaped cells emerge from the matrix, develop cytoplasmic fibres and become the cells of the cortex. Lying next to the cortex are found cells which after ascending become flattened and hyalinised to form the cuticle of the hair. The peripheral part of the follicle, called the inner sheath, consists of three concentric layers: (a) the cuticle, (b) Huxley's layer, and (c) Henle's layer. The inner root sheath cells are lost by scaling when they reach the orifice of the sebaceous gland and do not contribute to the emerging hair fibre.

From observations by electron microscopy, Parakkal & Matoltsy (1964) found that the multi-potential cells of the matrix develop structurally and chemically different products. The medullary and cuticular cells both develop amorphous cytoplasmic granules and form amorphous keratins, but have different sulphur contents. The cortical cells and the inner root sheath cells develop a 'filament–cement complex' containing different amounts of sulphur.

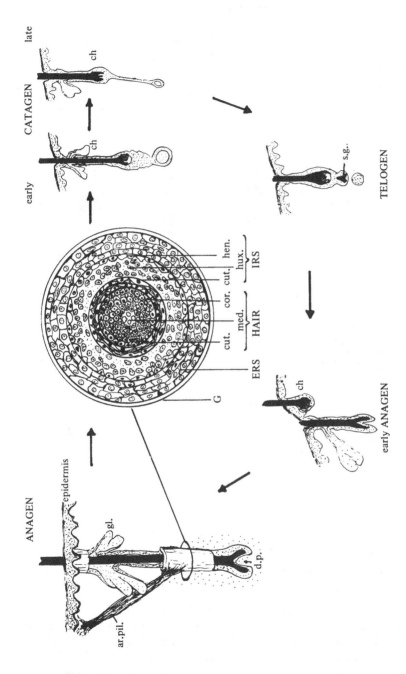

Fig. 15. Hair follicle and cycle of growth. ar. pil., arrector pili; ch, club hair; cor., cortex; cut., cuticle; ERS, external root sheath; G, glassy membrane; gl., sebaceous gland; hen., Henle's layer; hux., Huxley's layer; IRS, internal root sheath; med., medulla; s.g., secondary germ.

TABLE 2. *Differentiation and terminal products of the different cell types within the hair follicle*

	Inner root	Cuticle	Cortex	Medulla
Sulphur	low	high	high	low
Terminal product				
Differential product				

Matrix cell

Redrawn from Parakkal & Matoltsy (1964).

Whereas the horny substance of the inner root sheath cells is mainly produced by development of cytoplasmic filaments and trichohyalin granules, that of the cortical cells is formed through the production of cytoplasmic filaments and an amorphous interfibrillary substance (Table 2).

The cyclical activity of the hair follicle (Fig. 15), continues from its first formation during embryonic development until death of the individual, but the phases of the cycle and the quality and size of the hair present characteristics related more to the age of the individual. The first cycle extends from embryonic formation until 40 years after birth. The patterns of hair growth on the scalp from childhood through puberty to adulthood have been described in detail by Barman, Pecoraro, Astore and their colleagues in a series of papers from 1964 to 1979. The hair follicles during the early stages of development (from about the fifth month of gestation) are all in the growing active phase (anagen). Such uniformity of growth does not occur again during life except under certain pathological conditions. Between the tenth and eighth weeks before birth, some hair sites have already reached the intermediate (catagen) and resting (telogen) phase. These regions of the frontal and parietal scalp provide the first shedding events for the individual. The occipital region of the scalp provides an exception in that the hair remains in the anagen phase until after birth. Meanwhile, from about 6 weeks before birth, telogen hairs again appear in the frontal and parietal regions indicating a second cycle of hair growth.

In the majority of cases all hair goes into telogen phase immediately following birth, giving rise to the second episode of shedding. After this the phases are more irregular and eventually assume a mosaic pattern. Initially, the phases change suddenly and assume a wavelike characteristic progressing from the frontal to the occipital region. After about 18 weeks of life, hairs manifest either individually or in groups, a modification in the duration of the cycle, which gives rise to the mosaic pattern. The establishment of this new pattern introduces the life-long arrangement which undergoes only minor modification thereafter.

Differences related to skin complexion become less evident after the third month of post-natal life. At birth there is a marked difference between children with dark and fair hair. Those with dark hair are born with all hair in the anagen phase whereas fair-haired children show an increasing percentage in the telogen phase from occipital (20%) to frontal (50%) region. The number of hairs per square centimetre is generally greater in boys than in girls, but statistically the differences have not been shown to be significant on a population basis. When hairs are classified as thick, medium or thin, between 74 and 80% of hairs on the female scalp are found to be thick and between 76 and 83% on the male scalp. The medium and thin hairs each supply about 10%, although individual and seasonal variations are found. There is a statistical correlation between age and hair thickness in children.

The hair on the head of adults grows at the rate of about 0.3–0.4 mm per day, although the rates in women for cheeks, thigh, arms and abdomen have been measured at 0.2 mm per day (Barman, Pecoraro & Astore, 1964). Thickness variations across the normal adult scalp are not significant, but both thickness and density do change with age. On the average male scalp about 72% of hairs are 0.1 mm in diameter, 18%, 0.5 mm in diameter and

10%, 0.025 mm in diameter. The growth pattern (trichogram) of pubic hair showed changes in growth rate of 0.4 mm per day at the age of 15 years, decreasing to about 0.25 mm per day at 60 years (Astore, Pecoraro & Pecoraro, 1979). The density of hair also decreased with age from about 20 hairs per square centimetre at 15 years down to 10 hairs per square centimetre at 60 years.

Human hair cycles in the adult vary from site to site on the body surface. Saitoh, Uzuka & Sakamoto (1970) measured five different regions of the body and found great individual variation. The period of telogen for all regions including the scalp, varies from 8 to 19 weeks. Hair under the temple has a complete cycle of about 26 weeks, the arm 24 weeks, upper lip 19 weeks and the finger 18 weeks. These average values mask the variability at any one site, where the scatter of such measurements may lie between $\pm 30\%$ to $\pm 60\%$. By contrast the hair of the scalp may have a total cycle length of more than 300 weeks. It would take about 3 years for hair to grow 38 cm at a given steady rate, but most adults experience a difference in the growth cycle in summer and winter, with summer having slightly shorter cycles but growth rates increasing by about 12%. These fluctuations in growth rate reflect changes in cornification of skin with changes in ambient temperature. Except on the leg, all cycles tend to be longer in older people, who experience longer growth periods while resting periods remain unaltered. Growth rates in older people are slower.

Racial variations in hair are so conspicuous that the differences have been used as primary race characteristics. Today, the critical features concerned with differences in hair form are taken as: (i) average diameter, (ii) medullation, (iii) scale count, (iv) kinking, (v) average curvature, (vi) ratio of minimum to maximum curvature, (vii) crimp, and (viii) ratio of initial to straight lengths. The definitions of these terms and suggested methods for measurement have been given by Hrdy (1973). There is a high variance between populations and within individuals and the bulk appearance of hair is largely due to the shape characteristics such as curl, crimp and kinking.

The mechanical properties of hair are important from two aspects: (a) in grooming, and (b) in survival of hair. Study of mechanical properties not only permits the development of models for keratin structure and function, but allows the cosmetic chemist to consider the long-term effects of brushing, combing and hair care products. In an early study, Goldsmith & Baden (1970) measured the instantaneous Young's modulus by ultrasonic velocity methods during load-extension to break (fracture). The sonic modulus decreased from 117 N/m^2 to 88 N/m^2 as the humidity was increased from 6 to 53%. During extension of the hair at rates between 1 and 2.5% per minute, Goldsmith & Boden showed a decrease in elastic modulus over the first part of the yield region, but a rapid increase as the load approached the breaking stress. These data have since been confirmed by Henderson, Karg & O'Neill (1978) who also studied the initiation of fracture.

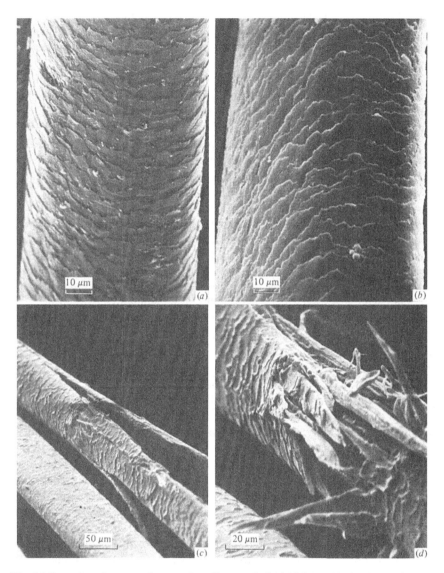

Fig. 16. Scanning electron micrographs of human hair. (*a*) Normal hair with damaged plates. (*b*) Normal hair, elliptical in section, with worn plate edges. (*c*) Simple longitudinal fracture of hair near ends. (*d*) Fracture and fragmentation of hair caused by bending and loading.

When extended in water the hair cuticle suffers multiple circumferential fractures with local separation of the cortex before the latter fractures uniformly across the fibre. By contrast, fracture in the dry state is more irregular with axial splitting of the cortex before or during splitting of the cuticle (Fig. 16). When examined by scanning electron microscopy, these dry fractures show great variability including cortical fibrillation. Under normal circumstances the onset of fracture is related to the wear on the hair cuticle from brushing and combing.

By measuring the number of cuticle cell layers on the hair shaft, it can be shown that there is an approximately linear relationship between distance along the shaft and wear. Various disorders of the hair, including monilethrix and pili torti, also lead to early fracture of the shaft (Fig. 16) (Dawber, 1977).

Various chemicals affect both hair growth and mechanical properties. The depilatory action of thallium, for example, has been known for almost a century. Accidental exposure to chemical agents, as well as therapeutic treatment for systemic disorders during the last two decades, suggests that a number of different mechanisms are concerned. Metals, such as thorium, barium and other elements normally appearing in hair as trace substances, are bound to the keratin fibrils and can lead to alopecia if present in sufficient quantity, the hair breaking off at the surface or within the follicle. Antimitotic agents (aminopterin or colchicine) can also cause alopecia, a relationship which may be overlooked since the drugs are frequently given in the treatment of neoplastic disease (Rook, 1965). The anticoagulants, coumarins and heparinoids, can induce alopecia, the effect being related to the maximum dose administered and not to any cumulative effect. In addition, dyes may cause fragility, change in colour or mask the distribution of natural pigment granules. All these factors have to be taken into account in the forensic description of hair (Porter & Fouweather, 1975).

Perhaps the substances most commonly associated with changes in hair are the steroids. The presence of testosterone in the circulation is responsible for changes in hair quality and the growth of post-pubertal hair. The continued presence of testosterone in the male leads to the typical male baldness pattern. Testosterone in the plasma reaches the hair follicles where it is converted into 5α-dihydrotestosterone, a tissue-active androgen, which is involved in the subsequent decrease in cyclic AMP. This in turn creates premature completion of the anagen phase. Years of repetition of this process probably leads to short vellus-type hairs and hair follicles and eventual baldness (Adachi *et al.*, 1970).

Diffuse hair loss in post-partum women is well known. The mechanism for this loss is not clear, but seems to be related to circulating hormone levels which sustain hair growth (Lynfield, 1960). Thus, after pregnancy there are many 'overaged' hairs which suddenly go into telogen. Post-partum defluvium is the restoration of normality rather than a pathological response. Nevertheless, it can be and often is psychologically very disturbing.

SEBACEOUS GLANDS

The sebaceous glands are relatively simple; they have partially divided saccules with relatively wide mouths and have little or no duct (Fig. 15). Around the periphery of the saccule a well-defined basement membrane is present. Between this and the epithelial cells no myo-epithelial cells are present, but there is a close relationship between the arrectores pilorum muscles and the gland. Elsewhere, for example in the nipple, bundles of smooth muscle can be found in the connective tissue between the saccules and their lobules.

Human sebaceous glands develop during the thirteenth to fifteenth weeks of gestation and attain quite a large size by the time of birth, but then shrink to comparatively small structures until the onset of puberty. With the onset of puberty and the stimulation by androgens, there is glandular hyperplasia and an increase in synthesis and secretion of sebum. The development of acne vulgaris during puberty is inextricably linked with and dependent upon the development of the gland. A full description of our understanding of sebaceous glands up to 1963 can be found in the review edited by Montagna, Ellis & Silver (1963).

The distribution of sebaceous glands over the skin surface largely follows the pattern of hair growth, but glands are large and numerous also on the midline of the back and largest on the facial disc, forehead, external auditory meatus and anogenital surfaces. There are about 400–900 large glands per square centimetre on the scalp, forehead, cheeks and chin. Over the rest of the body their frequency is less than 100 per square centimetre.

The major change in activity of the glands that occurs during puberty gives rise to an increase of more than sixfold in their rate of secretion from the age of 10 years to 18–19 years. Males secrete a little more than females throughout the post-pubertal period. Maximum secretion rate is about 5.5 mg lipid per 10 cm² per 3 hours, with a maximum average around 3.0 mg lipid. With increasing age the amount of sebum secreted falls steadily, until by the age 70–79 years men secrete at a mean level of 1.7 and women 0.9 mg lipid per 10 cm² per 3 hours (Pochi, Strauss & Downing, 1979). Administration of fluoxymesterone increases secretion rate in post-menopausal women.

Adrenal androgens probably play a large part in the control of sebum secretion. The exact route of control is however, not clear. Factors affecting secretion in man implicate the hypothalamus and pituitary gland (Burton, 1972; Ebling, 1972). Ebling has suggested that the response to steroids is mediated at the target site by one or more pituitary hormones, which permit the synthesis of 5 α-dihydrotestosterone and/or androstenedione, another intermediate of androgen metabolism. Both substances have a very marked effect on sebaceous secretion in hypophysectomised/castrated rats.

The role of sebaceous secretions on the surface of skin would appear to be twofold: the inhibition of water loss by means of a fatty layer and the

TABLE 3. *Regional counts of sweat glands*

Region	Average density per cm²: anatomical	No. active glands per cm²	
		European	African
Forehead	360	180	352
Cheek	320	—	—
Thorax	175	80	152
Back	—	64	—
Lumbar region	150	86	174
Abdomen	190	87	150
Thigh	120	62	98
Calf	150	—	—
Foot (dorsum)	250	—	—
Foot (sole)	620	—	—
Upper arm	150	81	115
Forearm	225	108	201
Palm	233	—	—

Compiled from Szabo (1962), Montagna & Parakkal (1974), Sarkany & Gaylarde (1968) and Sato & Dobson (1970).

provision of aromatic substances possibly associated with rudimentary sexual attraction. The frequent use of cosmetics and soap to remove or mask the effects of sebum on the skin tends to give superficial support to this latter suggestion.

SWEAT GLANDS

The ordinary eccrine sweat glands of skin are simple, unbranched, tubular, coiled glands, and are distributed throughout skin except on the nail bed, margins of the lips, glans penis and eardrums. They are most numerous on the volar surfaces of the hands and feet. The forehead and cheek have a gland density about half that of the volar surfaces but nearly twice that of the rest of the body. The density distribution is given in Table 3. Anatomically, inter-racial differences are small, although there are very large differences in the proportion of active glands at any one site.

The total number of glands in man is approximately 3 million. Each gland normally produces only a very small volume of fluid each day, but the total amount of sweat produced by an adult in a 24-hour period may be as much as 12 litres. Over shorter periods of time a rate of 3 litres per hour has been measured, a rate that exceeds the average ability to drink. During early foetal life the density of sweat glands is much higher than in the adult. A density of about 3000 per square centimetre has been measured in the thigh at 24 weeks of gestation; the number at birth is about half this (Szabo, 1962). By about 18 months the density has been reduced still further, by about two-thirds, to 500 per square centimetre. The adult density of about 120 per square centimetre does not change throughout the remainder of life. Similar data can be found for other body regions, the density changes reflecting the relative areal growth changes.

There are special sweat glands in the axilla, areola of the breast and circumanal region which are apocrine in type. In eccrine secretion, globules or granules pass through the cell wall without rupture, whereas in apocrine secretion the products collect in the luminal or distal part of the cell to bulge out to form a bleb. Some cytoplasmic elements are present, therefore, when the cell ruptures to release the secretory product. After release, the cell seals over and returns to its former elaborating phase.

Sweating mechanisms are extremely efficient, even though they do not enable man to survive in all environmental conditions. The patterns of sweating to be found are related to external and/or internal stimulation and are under the control of the nervous system. By electron microscopy many non-myelinated varicose axon profiles can be seen in the vicinity of both eccrine and apocrine glands (Uno, 1977). The glands are innervated with large numbers of cholinergic terminals and a few adrenergic terminals. As with the production of saliva, sweating can be a conditioned response.

Possibly because at least two types of innervation are possible, the physiological control of sweating remains uncertain. In experiments to determine the response to β-adrenergic stimulation, Warndorff (1971) gave isoprenaline to subjects with and without α-blocking agents. He suggested that the responses recorded were mediated by acetylcholine. In a later series of experiments Wolf & Maibach (1974) found that adrenaline induced palmar sweating and was not inhibited by atropine blockade or local anaesthetic. Emotional palmar sweating was completely blocked by atropine blockade, but spontaneous palmar sweating was unaffected. Palmar eccrine gland stimulation may be related to neither α- nor β-receptors, but rather to a previously undescribed type of adrenergic receptor.

Both noradrenaline and adrenaline can stimulate sweat glands, but sweat output is always less than that due to acetylcholine. Intradermal catecholamines give even smaller responses and glands fail to respond to repeated administration (Foster, Ginsburg & Weiner, 1970).

Since there is a wide variety of sources of stimulation of sweating, it is important to control as many of the variables as possible. These include: (*a*) high humidity, which may impair cooling, resulting in higher body temperature; (*b*) emotional and mental stimulation, which causes very rapid onset of sweating; (*c*) the position of the body during sweat collection; (*d*) local skin temperature, which may become highly significant when using methods to collect secretion from closely defined areas of the body; (*e*) metabolic rate, particularly with reference to time after last meal; (*f*) any therapeutic treatment or drug intake, particularly hormones; (*g*) sleep, when perspiration becomes highly erratic; (*h*) ratio of the sexes in mixed-sex test groups (females exhibit longer delay periods after stimulation before sweat appears at the body surface, even though over long periods there is little or no difference in the total amounts of sweat produced).

The patterns of sweating observed vary according to the methods used to record the output as well as with the methods used to stimulate the response. Primary stimulation may be achieved by the administration of drugs such as 0.5–1.0 mg acetyl salicylic acid followed by copious quantities of water or tea. Other drugs which induce sweating include acetylcholine, mecholyl, carbachol or pilocarpine. Exercise and application of local heat have both been used to provide primary stimulation, particularly when the subject is in a warm room or chamber. Studies on axillary eccrine sweating have been carried out under these latter conditions with careful control of temperature and humidity (Rebell & Kirk, 1962). After entering a warm atmosphere, however, there is usually a delay or latent period before sweating increases.

The variety of methods for recording sweat production indicates the difficulty in setting up a reliable and meaningful system. Colourimetric methods include the use of starch iodine as the indicator, or sodium chinazerin 2, 6-disulphide, or alcoholic cobalt chloride, or phenolphthalein, or bromophenol blue in silicone, together with many minor modifications of application and recording. For example, Spruit & Reynen (1972) used bromophenol blue as a powder with corn starch and gum tragacanth in the ratio 1:8:8 and dusted it onto the skin surface. Various galvanic methods have also been used (e.g. Allen, Armstrong & Roddie, 1973), as well as optical methods such as the absorption of infra-red (Uttley, 1972). Silver, Montagna & Karacan (1964) used a photographic method, in which the number and area of sweat spots were determined.

Even though the early attempts to summarise the reported data tended to be oversimplified, it is possible to suggest some general conclusions in relation to the distribution of sweating resulting from different groups of stimuli. Thermal sweating (Fig. 17) occurs over most of the body surface, but is at a maximum on the forehead, thoracic regions and the back. Emotional sweating depends upon the type of stimulus and its duration. In its simplest form it may be confined to the hands and feet, axilla and pubic regions, but emotional sweating induced by mental arithmetic in subjects at environmental temperatures between 26 and 29 °C increases dramatically not only on the hands and feet, but also on the trunk, arms and legs, and head and neck (Allen *et al*. 1973). Similarly, gustatory sweating may be associated only with the forehead (Champion, 1970), but more general sweating is commonly experienced although this may be due in part to emotional or temperature responses.

For a variety of reasons, some of which relate to cosmetic interests, studies on axillary eccrine sweating have been particularly detailed. From the work of Rebell & Kirk (1962) three different patterns of axillary sweating are now

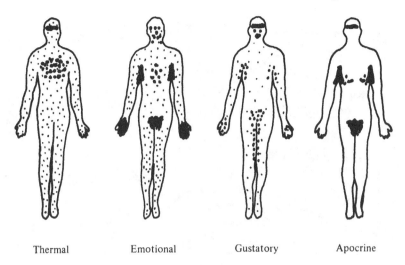

| Thermal | Emotional | Gustatory | Apocrine |

Fig. 17. Patterns of sweating.

identified, together with sub-groups. Briefly they can be summarised as: Type 1, active sweating extending beyond the boundaries of the axilla proper; Type 1 A, diffuse, even sweating throughout the axillary and peri-axillary regions; Type 1 B, profuse central axillary sweating, but with markedly reduced rates at the periphery; Type II, active sweating confined to the vault proper; Type II A, a pattern of sweating that is elliptical and fills the entire vault; Type II B, active sweating confined to a small region in the centre of the axilla; Type III, absence of sweating in the centre of the axilla, but with an active peripheral region. Each of these patterns is specific to particular individuals and consistent in time. The activities of the apocrine and sebaceous glands are not inter-related.

Not only have the patterns of sweating at the axilla been shown to be individual, but also the rates of sweating. Reller (1964) found three different types of response to thermogenic stimulation. In the first, there is a slow approach towards an equilibrium state where the rate of sweating is constant; in the second, there is a much more rapid approach towards equilibrium; in the third, there is a rapid increase in sweat production, which goes through a high maximum and decreases to a lower steady state. This final rate of perspiration approximates to that produced by other subjects showing the first or second type of response. All of the experiments carried out by Reller were at the relatively high ambient temperature of 38 °C.

Sweating has long been taken as a major body defence against temperature rise, one of the homeostatic systems. It is common experience that sudden changes in temperature induce increased sweat gland activity. Movement to a hotter climate induces higher sweat rates, but sometimes results in overheating of the body. The larger number of active sweat glands in Africans

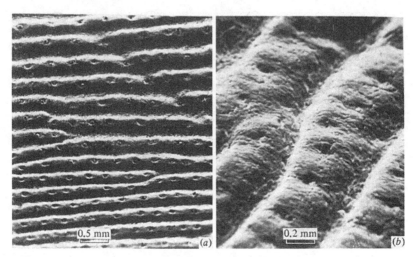

Fig. 18. Scanning electron micrographs of palmar skin to show sweat duct pores at the summit of ridges. (*a*) General low-power view. (*b*) At higher magnification the pores are seen as simple holes in the keratin outer layer.

was thought to contribute to their ability to live in high ambient temperatures (Montagna & Parakkal, 1974), but Peter & Wyndham (1966) have shown that while acclimatisation greatly reduces the number of inactive sweat glands, there is a decline in activity which accompanies prolonged exposure to hot, humid atmospheres and is attributable to glandular fatigue. The changes in gland activity are not related to increased body temperature, the degree of hydration of the skin or fatigue of the nervous system. Apparently, acclimatisation allows the body to reduce water loss while at the same time maintaining adequate body temperature.

In temperate climates, most sweat production is at the insensate level. The continued production of perspiration, however, has been seen as being important to the maintenance of skin. The duct mouth is present as a simple spiral channel through the epidermis with no special lining cells. The pores lie on the summit of the ridges of the palms and feet and at the junction of the primary creases in other regions of the body (Fig. 18). The duct proper begins at the epidermal–dermal junction. Diffusion of water from the spiral channel into and around the adjacent cells has been postulated as an explanation for the maintenance of the suppleness of skin at the folds, especially in winter and in other circumstances when the amount of sweat is low (Spruit & Reynen, 1972).

When pressure is applied to the skin, both the sweat rate and salt loss (sodium and potassium) are reduced. In order to explain this, Shuster (1963) proposed that the gland continues to secrete but that the fluid is resorbed by the ducts. The fact that under some conditions occlusion of the duct leads to rupture of the duct itself, has been taken as supporting evidence, but the ability to secrete against a hydrostatic pressure gradient has been known for much longer. Kuno (1934) had found that during moderate thermal sweating, arterial occlusion produced a gradual decline in the sweat rate after a latent period of about 10–15 minutes. Measurement of ionic concentrations of the surface sweat implies that there must also be resorption not only of water, but of electrolytes too. Thus, the concentration of electrolytes found in sweat is a property of the secreting cells of the coil and is not caused by differential absorption of water; but diffusion of water into and around the cells of the stratum corneum could alter surface electrolyte concentration at very low levels of secretion.

Spontaneous sweating in older subjects shows a reduction not only in the amounts of sweat secreted by each gland, but also in the number of active glands. Reaction to adrenaline is less, but the response to mecholyl is different in men and women: in older men there is a lowered response, but older women show no change when compared with young women (Silver, Montagna & Karacan, 1964). From these and other studies it is concluded that in old age two changes occur: (*a*) the intrinsic glandular condition, which deteriorates with age and limits the response to all types of stimuli, and (*b*) an extrinsic factor which affects sensitivity to cholinergic stimulation. The fact that there is apparently no detectable change in adrenergic response in older people might be expected since the adrenergic endings are much fewer in number.

3

THE SKIN IN DEPTH

Skin is a layered organ of which the *dermis* or *cutis vera* may be regarded as its primary structure. It is overlaid by the *epidermis* or *scarf of skin*, of five layers, the deepest of which, the *stratum germinativum*, is composed of soft protoplasmic, columnar cells, many in mitosis. The superficial layers of the epidermis, the *stratum corneum*, are hard and sometimes horny and vary greatly in number. The *hypodermis* is strictly not part of the skin although it carries the major blood vessels and nerves to the skin. It appears as a deep extension of the dermis and, depending on the region of the body and the state of nutrition, contains varying numbers of fat cells. In the abdomen, this layer may be 10 cm or more thick. The density and arrangement of the hypodermis determine the relative mobility of the skin.

The dermis is difficult to define since it merges with the underlying sub-cutaneous layer, the hypodermis. It has two layers, a *papillary* and a *reticular* layer (Table 4). The papillary layer is adjacent to the basement membrane of the epidermis and includes the ridges and papillae of fine connective tissue fibres. The sub-papillary plexus of vessels is a series of fine capillary loops. Some nerve endings are also present. By contrast, the reticular layer is the main fibrous bed of the dermis. It consists of thick, coarse, densely interlacing collagenous fibres parallel to the surface. Elastic fibres, prominent in youth, give tone to the dermis and permit tension even at low strain. The principal cellular elements of the dermis are the fibroblasts and macrophages, but fat cells, singly or in groups, may also be present. Pigmented branched connective tissue cells, *chromatophores*, can be found in certain areas, for example the nipple or circumanal region, where they do not elaborate their own pigment but obtain it from melanocytes. True dermal melanocytes are very rare.

Smooth muscle fibres are present as part of the hair structures, but are also scattered in considerable numbers in the nipple, areola of breast, penis, scrotum and parts of the perineum. It is contraction of these fibres that gives the skin in these regions its characteristic wrinkled appearance. Normally, striated muscle fibres terminate in delicate elastic networks of the dermis. It is this union between muscle, elastic fibre and the dermis that gives the face its intricate, voluntary movement.

TABLE 4. *The layers of the skin*

Epidermis	1. Stratum corneum	Thin, anucleate, squamous cells. Nearly all cytoplasmic structure lost. Thickest on palmar and plantar surfaces; thinnest on outer aspects of lips
	2. Stratum lucidum	Cell outlines indistinct, nuclei usually absent. Hyaline, band-like appearance due to presence of thick plates of keratohyalin. Absent in thin skin, most easily seen in glabrous regions.
	3. Stratum granulosum	Thin, irregular layer, two to four cells thick. Cytoplasmic granules of keratohyalin stain intensely with some acid and basic dyes.
	4. Stratum spinosum (rete mucosum or Malpighian layer)	Several layers, irregular polyhedral cells with prominent intercellular bridges. Cytoplasm basophilic (high RNA content). Basal layer sometimes columnar and exhibiting mitotic figures.
	5. Stratum germinativum	Single layer, columnar cells. Each cell has short processes which penetrates into basement membrane (thought to anchor cell). Mitotic figures frequent.
Dermis, or cutis vera	6. Papillary layer	Adjacent to basement membrane of epidermis, includes ridges and papillae of fine connective tissue fibres, loops of capillaries and some nerve endings.
	7. Reticular layer	Coarse, dense, interlacing collagen fibres, a few reticular fibres and numerous elastic fibres. Blood vessels, nerve endings, some sweat glands, hair follicles. Layers of collagen fibres not distinct in man.
Hypodermis	8. Adipose layer (sub-cutaneous layer, superficial fascia)	Deep extension of skin. Collagen fibres common to dermal fibres through branching and anastomoses. Varying number of fat cells, special nerve endings, sub-dermal plexus of blood vessels. Skin of eyelids, penis and scrotum devoid of fat.

THICKNESS OF SKIN

Attempts to measure skin thickness have been thwarted by the difficulties in defining the boundary between dermis and hypodermis. Many attempts have been made using fixed tissue, but even when correction factors for tissue shrinkage are applied, the values are only very approximate. Nevertheless, these studies on fixed tissue have allowed changes with age to be noted. More recently, techniques have been devised to measure skin thickness *in vivo* using non-invasive methods such as pulsed ultrasound or radiography. While there are no great changes in skin thickness demonstrable by these methods, it is possible to show that the thickness of male and female skin, when young, is significantly different, but because the natural scatter of the data increases with age, it is doubtful whether the differences above the age of 65 years are real. Indeed, large variability in the measurement of skin thickness has been a feature in older people. Table 5 shows some of the published data on fore-arm skin thickness as measured by various methods, but precise interpreta-tion of differences is not possible because of the varying numbers of subjects in the different groups and the different reliability levels between methods.

The measurement of epidermal thickness presents further difficulties, since the dermal papillae and rete ridges give an undulating lower surface. Again most attempts to measure thickness have been on fixed tissue in which shrinkage provides a large and uncontrollable variable. It is now well established, however, that epidermal thickness varies considerably over the whole body surface and variation at comparable sites between individuals is also high.

The studies of Whitton & Everall (1973) provide us with the most comprehensive set of data to date. To determine true epidermal thickness they applied Southwood's correction, which assumes that shrinkage in all dimensions is proportional to changes in specimen surface area. Different compensation factors were also applied to thick and thin skin to help to compensate for volume variability. The thinnest epidermis lies on the abdomen and thorax, where values from 16 to 50 μm have been reported. Skin is generally thicker on the arms and legs, and the plantar surfaces of the palms and the sole of the foot have the thickest epidermal layer. Mean values for epidermal thickness from a selected set of body sites are given in Table 6. Epidermal thickness on the *trunk* is less than 40 μm on more than 56% of subjects; on the *face*, 60% of subjects have an epidermal thickness less than 60 μm; on the *arms* and *legs*, only 49% of subjects have thickness less than 60 μm.

In a sophisticated study of epidermal tissue, Bergstresser, Pariser & Taylor (1978) obtained values for the epidermal thickness of skin from three body sites. The values obtained were: *volar forearm*, 35 μm; *upper back*, 37 μm; and *lateral thigh*, 40 μm. These values, however, did not include the thickness of the stratum corneum. If suitable values are added, such as those determined by Holbrook & Odland (1974), of *forearm*, 14.8 μm; *back*, 9.6 μm; and *thigh*,

TABLE 5. *Forearm skin thickness*

Sex of subjects	Age group	Type of measure	Range (mm)	Average skin thickness (mm)	Mean value[a] (mm)
Male	Under 65	X-ray[b]	1.0–1.7	1.3	
	Under 65	X-ray[c]	1.1–1.8	1.43	
	24–37	X-ray[d]	0.9–1.19	1.1	1.3
	24–37	Ultrasound[d]	1.0–1.16	1.12	
Female	Under 65	X-ray[b]	0.9–1.4	1.1	
	Under 65	X-ray[c]	1.0–1.7	1.34	
	28–37	X-ray[d]	0.82–0.95	0.88	1.26
	28–37	Ultrasound[d]	0.75–0.92	0.83	
Male	Over 65	X-ray[b]	0.7–1.2	0.9	
		X-ray[c]		1.19	1.1
Female	Over 65	X-ray[b]	0.6–1.2	0.9	
		X-ray[c]		1.06	1.0

[a] Mean weighted by subject number tested.
[b] Black (1969). Data less than 2% s.e. under 65, less than 5% over 65.
[c] Sheppard & Meema (1967). Data less than 3% s.e. under 65, less than 20% over 65.
[d] Alexander & Miller (1979). Data less than 2% s.e.

TABLE 6. *Mean values for full epidermal thickness*

Body site	Mean thickness (μm)
Palm	429[a]
Fingertip	369.0
Back of hand	84.5
Forearm	60.9[b]
Upper arm	43.9
Thoracic region	37.6
Abdomen	46.6
Upper back	43.4[b]
Lower back	43.2
Thigh	54.3[b]
Calf	74.9
Forehead	50.3
Cheek	38.8

All values are from Whitton & Everall (1973), except: [a] estimated from published data; [b] also from Bergstresser *et al.* (1978).

10.9 μm, they obtained a striking agreement with the values obtained by Whitton & Everall.

In everyday life there is a balance between the production of new cells, their migration and the rate of loss of cells from the surface. Adjustments are made to meet changing patterns of events from within the body or the effects of the environment, but the process is not instantaneous and this accounts for the considerable variation in thickness encountered. Mechanical stresses affect the epidermal thickness within a period of only 2 or 3 days. Squier (1980) measured the effect of implanting small springs in the back-skin of hairless mice. He found that the thickness of the Malpighian layer (the combined stratum germinativum and stratum spinosum) increased from 17 to 38 μm over a period of 4 days. Control animals showed only very small changes in thickness during this period, even the effects of a sham operation lasting only 24 hours.

EPIDERMAL GROWTH

'Regeneration time', 'renewal time', 'replacement time' or 'turnover time' are all synonyms for the *average* time for a cell population to reproduce itself. There is a measurable regeneration time for the germinative layer of the epidermis. Cells transfer or migrate from this basal layer through the other layers until desquamated from the surface. Because of this movement of the keratinocytes towards the surface, the regeneration time for the other layers in the epidermis, which do not exhibit mitotic activity, is dependent on the regeneration time of the basal layer. The time taken for migration through all the layers of the epidermis is the transit time.

The measurement of the Mitotic Index (MI), or the rate of mitosis, is not always clearly expressed in the literature. Usually the MI is given as the number of cells seen in one or other stage of division per thousand viable cells, but this may also be expressed as a percentage. A variety of drugs can be given to arrest mitosis and so facilitate the count. Using Colcemid (*N*-deacetyl-*N*-methylcolchicine), Fisher (1968a) showed the values for the MI 5 hours after administration of the drug, (sometimes called the 5-Hour Index), to be 1.44 to 1.17. He also found that the mitotic rate varied throughout the day. This diurnal rhythm gave rise to MI values ranging from 0.1 to 0.9, the lower figures relating to a different arrest time for this experiment. From comparison of the average MI at different times during the day – during rest periods, at meal times, and exercise – he concluded that the diurnal variation was due to fluctuation in the corticosteroid hormone levels.

To calculate the duration of mitosis, the formula of Hooper (1961) is frequently used. From data from tissue with and without colchicine, the value of T_m (the duration of mitosis) is given by:

$$T_m = \frac{\% \text{ mitotic figures at stage S}}{\% \text{ mitotic figures at metaphase}} \times t$$
$$\text{(after treatment with colchicine)}$$

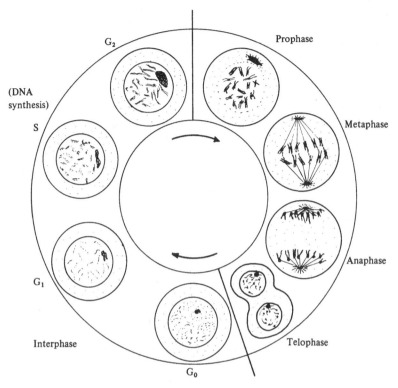

Fig. 19. Diagrammatic representation of the cell mitotic cycle. The four phases of interphase include the process of DNA replication.

where t is the time in minutes from colchicine treatment to fixation of cells (or death) and S is the stage of DNA synthesis, prior to prophase. (A general scheme for the stages of mitosis is given in Fig. 19.)

Fisher (1968b) calculated the duration of mitosis for normal epidermal keratinocytes as approximately 90 minutes, which agrees with the later *in vitro* estimates of Flaxman & Chopra (1972) and Chopra & Flaxman (1974). On the other hand, Weinstein & Frost (1969) reported a T_m value of 60 minutes. The total time for mitosis is probably variable within certain limits and is influenced by delays in the G_1 and G_2 phases (Fig. 19). Factors affecting the duration of mitosis are known to include environmental temperature, but the precise description of all contributing elements awaits elucidation.

From a knowledge of MI and the duration of mitosis, it is possible to estimate the regeneration time for the tissue. Epidermis, however, is a layered tissue with ageing cells and so it is necessary to know the transit times for cells moving towards the surface. This could be complicated further if cells can also move laterally. Fortunately, the discovery of a prescribed structure which defines the lateral extent of the proliferating unit in animals, has led to a reassessment of regeneration time values. With such a unit of structure, all that it is necessary to know in addition to MI and duration of mitosis, are the number of cell layers, cell density and desquamation rates from the surface of skin.

THE EPIDERMAL PROLIFERATING UNIT, EPU

Most of the early estimates of regeneration time were based on the assumption that the basal or germinative layer has a homogenous proliferative cell population. This is an oversimplification. The epidermis in many species is known to be highly organised with cornified cells arranged in precise columns. The columns are arranged above groups of basal cells from which the keratinocytes migrate. This discrete arrangement of cells, has been termed the epidermal proliferating unit (EPU).

An EPU (Fig. 20) contains a number of basal cells together with two or more post-mitotic cells and a Langerhans cell. This group of cells may occupy more than one layer. Melanocytes are scattered irregularly along the epidermal basal layer but are not present in every EPU. Above this group of cells rest a number of nucleated flattened cells, overlaid by several cornified 'dead' cells. The precise numbers depend upon the site. Recognition of the limits of the unit is restricted to regions with low, steady-state kinetics.

The basal layer contains non-keratinocyte-producing cells (melanocytes, Merkel cells, etc.) as well as cells temporarily in a period of quiescence (G_0). Many cells may have different inherent potentials for division; some will be in the post-mitotic maturing stage, typical of the migrating cell. The small number of remaining cells will have an unlimited potential for division. This last group has been called the sub-population of stem cells. Each EPU has a daily cycle of cell production with each primary stem cell having approximately a weekly cycle.

From the primary stem cell a new intermediate cell is produced with a potential to divide again up to three times (Potten, 1975). The two daughter cells (Fig. 20b, cells numbered 2 and 3) have different maturities and the first two divisions have cycles of about 5 and 8 days, the subsequent daughter cells being pushed to the peripheral positions where they could divide again. Thus, in this model, there is 1 primary stem cell to 5–7 dividing maturing cells and 2–3 post-mitotic ageing keratinocytes. It follows that each primary stem cell divides once a week, but within the unit as a whole, a cell is produced each day.

Using this type of model, we see that estimation of regeneration time does not depend upon the concept of the EPU itself, although it does provide a framework for understanding the process. By identifying the corneocyte and keratinocyte compartments, Bergstresser & Taylor (1977) used published data to calculate the regeneration times for epidermis. They took the rate of corneocyte desquamation as 20 layers in 14 days, or 1.4 cell layers per day.

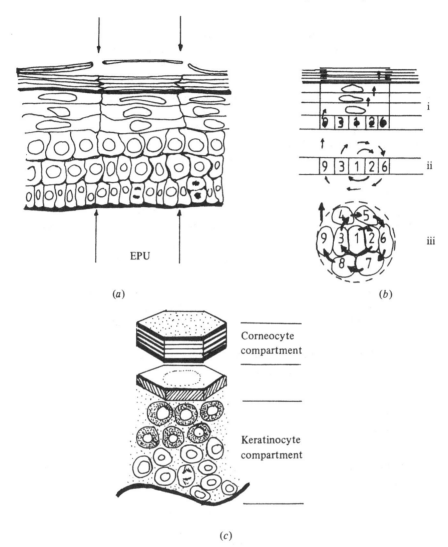

Fig. 20. The epidermal proliferative unit, EPU. (*a*) Section through skin showing an EPU. (*b*) The movement of cells from mitosis to corneocyte formation. (*c*) The concept of the corneocyte and keratinocyte compartments used in the calculation of turnover time.

The concept of the EPU is based upon the identity of the unit under a single corneocyte: thus, for one EPU, the desquamation rate is 1 per day. The number of keratinocytes under one surface corneocyte is given by the product of the number of keratinocytes per unit area (4.7×10^4/mm²) and the surface area of the corneocyte (9.4×10^{-4} mm²) and is equal to 44. The regeneration time for the keratinocyte compartment is then calculated from the rate of maturation (1.4 cells per day) and the number of keratinocytes, viz. $44 \div 1.4 = 31$ days. If this is added to the regeneration time for the corneocytes, 14 days, the regeneration time for the entire epidermis is 45 days.

Using earlier models, Halprin (1972) had calculated the regeneration time to be 52–75 days. Cells of the germinative layer had a regeneration time of between 12 and 19 days, the Malpighian layer a transit time of 26–42 days and the stratum corneum of about 14 days. Other results have reported values for the total regeneration time ranging from 28 days to 56 days. By contrast, psoriatic skin has a regeneration time of 8–10 days, made up of germinative layer 1.5 days, Malpighian layer 4–6 days and the stratum corneum 2 days.

While the model for the EPU appears to hold for many animal species, it is seldom confirmed for human skin. Bergstresser & Chapman (1980) reported studies on this apparent anomaly and concluded that the formation of columns of cells is related to the rate of cell production. In normal human skin the rate is assumed to be too high to allow regular stacking of cells, the high rate of cell production being stimulated by washing, work abrasion and other normal activity. Stacking can be found, however, in regions of skin taken from grossly undernourished patients. In those animals which readily display stacked epidermal cells, accelerated mitotic activity, induced by mechanical injury, results in a loss of corneocyte column formation. Thus, regular stacking of epidermal cells occurs only when certain criteria are met and the most important of these is a low mitotic index.

CHALONES

During the waking, active period of the organism, many cells progress towards division, but do not complete the cycle. They are prevented from proceeding further. Within a few hours of the onset of sleep, all the accumulated cells have completed the mitotic cycle and the rate of mitosis falls, but remains higher than during the waking period.

It was from a study of cell proliferation in epidermis that Bullough & Laurence (1960) introduced the concept of a chemical inhibitor of mitosis. The study of increased mitotic activity that accompanies any local mechanical damage leads to the following conclusions:

    (a)   that it is not caused by hyperaemia;
    (b)   that a mitosis-stimulating 'wound hormone' is not involved;
    (c)   that it is due to the loss of a mitotic inhibitor, present in normal epidermis;
    (d)   that the inhibitor (called the epidermal chalone) is tissue-specific, in both origin and action.

The epidermal chalones are glycoproteins of about 30 000 molecular weight (Boldingh & Laurence, 1968) which can pass from cell to cell through the dermis and are present in urine. During stress and unusual activity mitosis almost ceases, but, following adrenalectomy, it is found to rise to a high level and the diurnal rhythm ceases. Chalone action is related in some way to adrenaline, but adrenaline is not itself a mitotic inhibitor.

Chalones act at two stages in epidermal growth: (*a*) in the basal layer to prevent the mitotic cell from entering the division phase, and (*b*) in the outer region of the stratum spinosum to prevent cells from completing the process of keratinisation. In wounded or psoriatic tissue, chalone concentration falls, with an associated increase in mitotic rate and a compensating increase in post-mitotic death. Thus, during irritation or damage, there is no major increase in epidermal thickness nor is there a great thinning of the epidermis during decreased mitotic activity. This is an important safety mechanism. It does not, however, of itself, explain why epidermis varies in thickness from one body site to another.

Two epidermal chalones, Epidermal G_1 chalone (Marks, 1975) and Epidermal G_2 chalone (Boldingh & Laurence, 1968), have been isolated and in their highly purified form have been shown to be specific to one phase of the cell cycle (Thornley & Laurence, 1976) (Fig. 19). In each body site, each cell reproduction phase lasts for a different time and, in order to investigate chalone effects fully, care has to be taken to select suitable contrasting regions. Laurence, Spargo & Thornley (1979) used pinna epidermis, which has an S-phase duration of 18.4 hours, and sebaceous gland tissue, in which S-phase lasts 18.8 hours, to obtain more accurate data on tissue specificity of chalones. By extrapolating their results with respect to time of action, they demonstrated that the G_1 chalone acts at the point of entry into DNA synthesis, and that the duration of S-phase remains largely unaltered. This result was contrary to earlier reports that G_1 chalone acts some hours before entry into the S-phase, but the earlier work did not take into account the effects of short S-phase time and diurnal rhythm. It would appear then that the chalone acts as an inhibitor of a specific biochemical process and so controls the cycle of cell division.

Such a hypothesis would be more tenable if we understood how the action of chalones at the site of intervention proceeded. The relationship between the activity of chalones and that of adrenaline could be an important clue. Indeed, Bullough & Laurence (1964) have suggested that it is a complex between chalone and adrenaline that forms the actual inhibitor. Chalones themselves do have some similarities with hormones with respect to the control of proliferation and differentiation of cells. Extrapolating current knowledge of the mode of action of hormones, Iverson (1969) went on to postulate that chalone effects may be mediated through the cyclic AMP (cyclic adenosine 3′,5′-monophosphate) system in a manner similar to that of adrenaline or isoproterenol. The chalone would interact with adenylate cyclase to stimulate the enzyme to produce sufficiently high levels of cyclic AMP to maintain the ratio of cyclic AMP to cyclic ADP, thus ensuring epidermal homeostasis. Vorhees & Duell (1971) used this suggestion to derive an explanation for the proliferative skin disease, psoriasis. In support of their arguments, it was known that adrenergic substances, such as isoproterenol, produce stimulation of the adenylate cyclase within the epidermis and that there is an increase in epidermal cyclic AMP before mitotic inhibition. In addition, Bronstad, Elgjo & Oye (1971) found that the concentration of adrenaline necessary to stimulate hamster skin was much higher than in specific target tissues and claimed that this implied the presence of another hormone of glandular or local origin acting as the physiological activator. This 'local hormone' would now be identified as a chalone.

The control of cell mitosis and hence tissue regeneration and growth by chalones or hormone-like substances within specific tissues, is now largely accepted. But the production of chalones remains a mystery. Within the epidermis there are a number of different cell types in addition to the keratinocytes. Melanocytes and supra-basal dendritic Langerhans cells are but two of the many cells present. The role of melanocytes has already been described and their principal function identified. The Langerhans cells, on the other hand, have been the subject of much speculation: are they sensitive cells related to the nervous system, or are they specialised cells giving support to the keratinocytes?

LANGERHANS CELLS

After a variety of staining procedures, epithelial cells *not* of the keratinising type can be found in sections of skin. We now know that the so-called clear cells (Masson, 1926), which fail to exhibit cytoplasmic staining with haematoxylin and eosin, include the branched cells (Merkel cells, Langerhans cells and melanocytes) as well as the unbranched lymphocytes. Of these, the Langerhans cell can be distinguished by its characteristic morphological features: an elongated, branched cell with polymorphic nucleus and ATPase-positive cytoplasm containing Birkbeck granules, but no desmosomes.

Langerhans cells are found throughout the stratum spinosum, but not beyond, whereas Merkel cells and melanocytes are strictly confined to the basal layer. How the Langerhans cells maintain their position against the movement of the keratinocytes is not understood. It is conceivable that the shape of cell is important to the maintenance of position or that the cell's morphology changes if it passes into the upper layers.

The origin and function of the Langerhans cells have been a puzzle for a very long time. Langerhans (1868) thought the cells were probably neural in origin, because of their shape, and hence migratory. Others have since suggested that they were melanocytes – either worn out or dividing or arrested – and therefore derived from the mesoderm. Even suggestions associating the cells with ectoderm can be found. Perhaps the strongest morphological evidence for the migratory origin of the cells lies in their distribution throughout the epidermis. In many mammals the commonly used experimental sites (e.g. ear, back and abdomen) have Langerhans cell densities ranging from 500 to 1500 cells/mm^2, whereas in other sites (e.g. hamster cheek) there may be as few as 200 cells/mm^2. Although detailed descriptions of the distribution in human epidermis do not exist, it is assumed that the magnitude of the density differences is of the same order. There is no record of Langerhans cells in mitosis and so the assumption of migratory activity is commonly held.

Silvers (1957) undertook the first definitive attempt to identify the origin of the cells. He transplanted neural-crest-free limb buds onto spleen and then looked for Langerhans cells in the epidermis. The presence of aureophilic (gold-staining) dendritic cells in the epidermis, he suggested, ruled out the possibility that Langerhans cells were of neural origin, and hence related to the melanocyte. There are many criticisms that can be levelled against the experimental procedure he adopted, not the least being that spleen itself contains Langerhans-type cells. Nevertheless, it is intriguing to discover that his conclusions and the later ones of Breathnach *et al.* (1968) have since been upheld.

The most recent suggestion is that the Langerhans cell can be derived from bone marrow (Tamaki, Stingl & Katz, 1980). The body, and in particular the bone marrow, contains a pool of precursor cells some of which may develop into Langerhans cells – which are therefore related to the monocyte–macrophage–histiocyte series.

In studies *in vitro*, Langerhans cells have been shown to have unique immuno-responses in the epidermis and can replace macrophages in their capacity to induce antigen-specific and allogenic T cell activation. Thus, some support can be found for the suggestion that Langerhans cells form a sub-population of 'macrophages' (Thorbecke, Silberberg-Sinakin & Flotte, 1980; Green *et al.*, 1980). On this basis, the cells may function in antigen presentation, lymphokine production, provision of microenvironment for T lymphocytes and prostaglandin secretion. If an epidermal suspension rich in Langerhans cells is pulsed with antigen, proliferative responses in immune cells can be induced which are of the same magnitude as those induced by antigen-pulsed macrophages (Toews, Bergstresser & Streilen, 1980). In other words, the Langerhans cells may form the first active line of defence for the body.

The clinical implications are important. Such a function would impose a specific role on Langerhans cells in the provision of sensitising factors in contact hypersensitivity and skin graft rejection. (For a development of these ideas see Stingl *et al.*, 1980.) But the complete functions of the Langerhans cell may be even more complex than this.

As well as being involved in contact allergy and immunologic reactions, the cell may also be involved in the production of inhibitors of epidermal cell proliferation (Potten & Allen, 1976). The Birkbeck granules may be involved in the production of chalones or chalone precursor substances. However, attempts to correlate Langerhans cell density with proliferation rate have not yielded conclusive results, even though their number is reduced in diseases characterised by parakeratotic differentiation (chronic psoriasis and chronic eczema) and increased in diseases with ortho-keratinisation (lichen ruber and icthyosis).

In experiments on animals, Langerhans cells have been reported in the 'scale region' of rat tail epidermis after vitamin A treatment (Schweizer & Marks, 1977), but, on the other hand, it is known that vitamin A deficiency results in conversion of non-keratinised bladder epithelium to an ortho-keratinised squamous epithelium accompanied by the appearance *de novo* of Langerhans cells. At best, these experiments have yielded equivocal results and the relationship between vitamin A, adrenaline and chalones remains speculative. Nevertheless, recent advances in knowledge of Langerhans cell function makes this currently one of the more exciting fields of discovery in skin.

THE DERMO-EPIDERMAL JUNCTION

The junction between epidermis and dermis serves three primary functions: (1) epidermal–dermal adherence; (2) mechanical support for the epidermis; and (3) a partial barrier to the exchange of cells and some large molecules. By light microscopy, the junction is seen as a feltwork of reticular (silver-positive) fibres, through which some thin elastin fibres penetrate. The feltwork also contains antigenic materials which interact with basement membrane antibodies.

By electron microscopy, the junction is seen to contain four structural components: (*a*) basal cell plasma membrane with its attachment devices (hemidesmosomes); (*b*) the lamina lucida (or inter-membranous space); (*c*) the basal lamina (basal or basement membrane); and (*d*) the sub-lamina fibrous elements. The latter fibrous lamina contains the anchoring fibrils and some bundles of microfibrils as well as single collagen fibrils (Fig. 21).

Dyes such as Methylene Blue and Evan's Blue do not cross the junction. Unless damaged, therefore, cells and large molecules do not passively diffuse through the barrier, but glucose and salt ions can pass through freely. The process of transfer is not thought to be controlled through electric charge but cellular secretions may play an important part, since pretreatment with hyaluronidase increases the amount of substance that can be passed and chondroitinase and trypsin both lead to breakdown or rupture of the membrane. Okubo & Sano (1973) suggested that the basal lamina can alter its physical properties and hence its passive permeability characteristics through the conversion gel to sol, and that substances such as the proteinase–hyaluronidase system may play a part in changing the membrane at local or focal points.

The concept of a gel to sol conversion raises the question of interpretation of the structures seen by electron microscopy. It is well known that fixation, dehydration and subsequent embedding procedures change both the overall

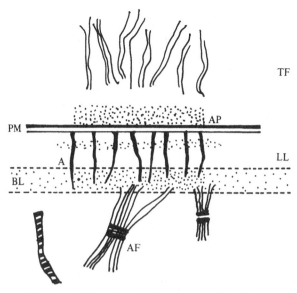

Fig. 21. Diagram of dermal–epidermal junction to show hemidesmosomes, lamina lucida (LL) and basal lamina with the sub-lamina fibrous elements. Tonofilaments (TF), attachment plaque (AP), and plasma membrane (PM), make up the hemi-desmosome. Anchoring filaments (A), basal lamina (BL) and anchoring fibrils (AF) are also shown.

dimensions of the structural elements and also the dimensions of the 'clear spaces' (lamina lucida). It is likely that in living tissue the lamina lucida is filled by the hydrated basal lamina. Nevertheless, the changes in these layers recorded in disease states must reflect important alterations to both the physical and functional condition of the lamina.

The sub-lamina elements should play some part, as yet undetermined, in the controlling mechanism, in addition to their structural binding properties. There are three types of fibre present – oxytalin, elaunin and collagen – and these are all formed from tropocollagen (Rodrigo & Cotta-Pereira, 1979). The continuity of the oxytalin fibres with the elaunin fibres has been demonstrated by electron microscopy of the junction region, and this similarity with the findings in normal elastogenesis leads to the speculation that the junction fibres have a relatively high degree of elasticity, a property useful if the boundary is to resist shear forces. Electron microscopy also shows fusion of the dermo-epidermal basement membrane lamina with the basement lamina of thin superficial nerves (Briggaman & Wheeler, 1975), but *not* with the external collagenous lamina of the blood vessels. This is taken to imply a difference in the composition of the basement membranes of nerves and blood vessels of this region, which may be reflected in the functional and structural arrangements observed.

Proof of the concept of 'anchoring fibrils' is somewhat tenuous. In epidermolysis bullosa, where lesions occur within the lamina of the junction, the skin lacks these 'anchoring fibrils'. The need for such fibres to provide mechanical stability to the junction zone is thereby given some support. The larger elastic and collagen fibres of the reticular dermis are not linked directly to the junction zone and except through the ground substance could not provide the required mechanical support. In addition, the size and frequency of the capillary loops, and the delicate nerve endings, require the soft, pliable supporting structure provided by the gel-like matrix. Thus, the continued integrity of the papillary tissue, which is necessary for the well-being of skin, appears to be provided by delicate connective tissue arrangements.

THE DERMIS

The dermis in lower-order animals has a relatively high degree of fibre orientation (Fig. 22), but in man there is no specific orientation or layering of tissue. On first examination the dermis appears to consist entirely of random bundles of collagen fibres, which in older people have aggregated together in a partially layered or heterogenous mass of tissue. On closer inspection it can be seen that the collagen bundles lie approximately parallel to the surface and when extended tend to orientate in the direction of the applied load. This tendency towards directional orientation has been described in terms of Langer's lines and similar phenomena. Langer (1861) showed that when a round awl is inserted into cadaveric skin, the resulting hole was ellipsoidal not round. By joining the long axes of these ellipses together, lines were formed across the surface of the body representing lines of tension in skin. The phenomena of skin tension, first described in detail by Langer, are discussed at length later.

COLLAGEN

Collagen is by volume the principal constituent of the dermis, but throughout the body can be found with various compositions and in different structural aggregations. In skin there are at least three forms of collagen. During periods of accelerated growth, for example during foetal life, the type III collagen predominates (Table 7), while the main constituent of normal adult skin is type I. The differences between collagen types lie not only in the amounts and distribution of amino acids along the collagen molecule, but also in the degree of hydration. Dermal collagen tends to have less water associated with it than do other collagen types.

The distribution of collagen in skin is not uniform. There is in young adult skin an enrichment of type III collagen in the papillary layers, while the deeper reticular dermis consists predominantly of type I collagen. This has led to the suggestion that collagen is formed in the papillary layer and migrates into the reticular layer where it ages before being resorbed, but there is, as yet,

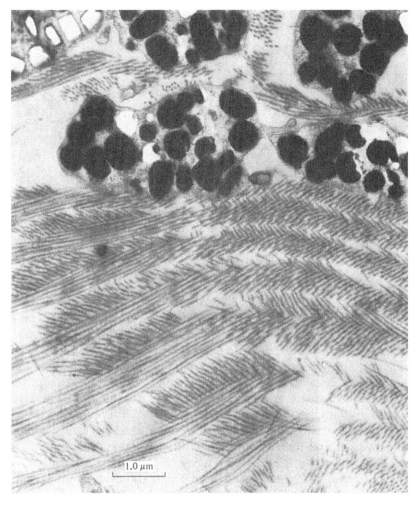

Fig. 22. Transmission electron micrograph of dermal tissue from toad showing highly ordered arrangement of collagen bundles.

TABLE 7. *Collagen composition and water content*

	Biochemical composition (residues per 1000)						Hydration	
Type	Glycine	Proline	OH-Proline	Lysine	OH-Lysine	Cystine	(%)	Site
I	330	129–135	90–96	30.32	5	—	0.5	Reticular dermis
II	320	108	13	21	—	10.0		Cartilage
III	348–360	107–109	112–121	30–35	5	3	0.5	Foetal and adult papillary dermis
IV	325	60	163	5	42	8	12.0	Basement membrane

From Meigel & Weber (1976) and Uitto & Lichtenstein (1976).

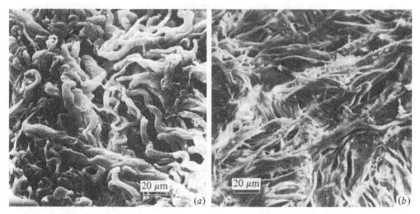

Fig. 23. Scanning electron micrographs of dermal collagen from (*a*) a child, 3 months old, and (*b*) an adult, aged 45 yrs.

no satisfactory evidence to support the movement of collagen into the deeper dermal regions. Indeed, two factors would tend to suggest that collagen is formed *in situ*: (i) the collagen fibres are long, convoluted, cross-linked structures, and (ii) fibroblasts can be found at all levels throughout the dermis.

Even on superficial inspection it is evident that the quality of skin changes with age, and it is not surprising to find that the appearance of the dermal collagen changes too. In young skin the fibres are distinct, convoluted and appear to be interwoven (Fig. 23*a*), whereas in older skin collagen takes on a more sheet-like aspect, appears more dense and is less flexible (Fig. 23*b*). It is not only with age, however, that skin changes. Even in young people there are marked changes in dermal structure in sun-damaged areas, and this, together with other information, has led to the suggestion that the course of ageing is a continual process which responds to environmental changes.

The amount of collagen extractable from skin also changes with age, to such an extent that the method has been used to identify the age of a person. Samples of dermis are homogenised in citric acid and the soluble part precipitated 48 hours later (Bakermann, 1964). The relationship with age is logarithmic and the following formula is used:

$$N = N_0 \cdot e^{-A/k}$$

where $N_0 = 1.3$ mg/g wet dermis
A = age of subject
k = age constant (23.5 years)
N = extractable collagen

At puberty about half the total collagen can be recovered as soluble collagen, whereas at menopause (or about 45 years) the recoverable collagen is only one-seventh, and at 60 years is usually less than a fifteenth.

Collagen can also be prepared from skin by making neutral salt or pepsin

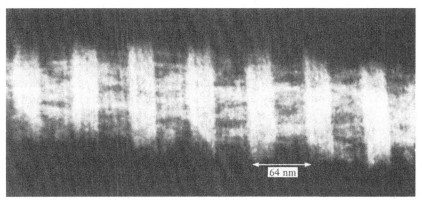

Fig. 24. Transmission electron micrograph of collagen fibril negatively stained with phosphotungstic acid, with a repeat periodicity of about 64 nm.

extracts. The extracts can be dialysed in 0.225 M citrate buffer (pH 3.7) against tap water and reconstituted fibres prepared by stretching the resulting gel. Collagen fibres including natural fibres have a characteristic repeating period of about 64 nm. By X-ray diffraction it can be determined that the periodicity for air-dried specimens is near 64 nm and for wet tissues near 67 nm. This periodicity has been interpreted in terms of an axial staggering by 67 nm of the 300-nm-long triple-helical collagen molecules within the fibrils. In the electron microscope the banding seen at high magnification represents the arrangements of amino acids along the fibril (Fig. 24), and comparisons of collagen from different sources reveal observable differences in banding pattern. There is little or no change in the measured periodicity between relaxed specimens and those under tension (Brodsky, Eikenberry & Cassidy, 1980), despite earlier reports from dried specimens which showed an increase with tension, and those of Nemetschek *et al.* (1978) who reported an increase in periodicity in wet specimens under very high tension. In skin the periodicity measured (65.2 nm) appears to be smaller than that for tendon (67.9 nm) and discussion centres on the relative composition of the two types of collagen. The skin, while being primarily type I, contains a significant amount of type III collagen whereas in tendon type III collagen has never been reported as present. Type I collagen from skin exhibits the same periodicity as type I collagen from tendon and therefore is not of itself leading to the shorter periodicity in skin. Reconstituted type III fibres have, if anything, a slightly longer spacing than type I fibres, so an interaction leading to a shorter form seems unlikely. At present, it is suggested that there is some particular environmental factor associated with the dermis that leads to the shorter form. Differences in water content have been excluded, since the intermolecular equatorial spacing, which reflects hydration state, is in both skin and tendon at the same position, near 1.48 nm. It is possible that the proteoglycans in skin reduce the charge repulsion and allow collagen to adopt a less extended form.

It is popularly assumed that dermal structures degenerate with age. Common manifestations are fragility, poor wound healing and wrinkling, and it is difficult to envisage how these would occur without involving collagen. Collagen is, however, remarkably resistant to degradation by most proteases; specific enzymes, collagenases, are necessary for the initiation of breakdown. It must, therefore, be under specific control mechanisms regulating its rate of both synthesis and breakdown. Young collagen is known to be more digestible by collagenase than more mature collagen (Kohn & Rollerson, 1960), yet there is a measurable decrease in total collagen present in skin with age (Shuster & Bottoms, 1963). Sex is one of the factors known to be of significance. Thigh skin loses collagen at a faster rate in ageing women than in men; the difference is even greater in the forearm, where the effect of solar radiation is an additional contributing factor.

The basic molecular structure of collagen seems to remain into old age even though the quantity is reduced. There are, however, some diseases where major changes in collagen structure have been noted. There are broad-banded collagen fibres in the granulatomatous areas of skin biopsies from lepromatous patients (Edwards, 1975). Similar broad-banded fibres (100–120 nm) have been reported from time to time in normal tissues such as Decemet's membrane, Meissner's corpuscle and the trabecular region of human eye, but always as an exceptional observation. They appear to be more common in the connective tissue associated with tumours, neurofibromas and neoplasms. Under experimental conditions, Robertson (1964) found that many factors modified the morphology of native collagen *in vivo*, including vitamin C deficiency, mucoproteins, polysaccharides and lipids as well as endocrines. The quality of the collagen remaining in old age could depend quite markedly on its association with tissue proteoglycans.

ELASTIC FIBRES

The elastic connective tissue fibres are highly branching structures, which in skin may aggregate to form very large bundles, notably in the deeper dermal regions of the back. Electron microscope examination of elastic connective tissues indicates that the fibres consist of two distinct components (Ross, 1973). The major component has an amorphous appearance with no distinct periodicity. This represents the elastin protein. The fibre has an envelope of peripheral microfibrils, which in transverse section appear tubular and are of diameter 10–12 nm. These two proteins, the elastic and microfibrillar components, exist in various connective tissues.

The basic molecular unit of elastin is a linear polypeptide with an approximate molecular weight of 72 000 daltons. It has a highly characteristic amino acid sequence and lacks tryptophan, cystine and methionine. About one-third of the total number of amino acids is glycine, which is randomly distributed along the polypeptide chain, unlike the even distribution typical of collagen. Instead, multiple sequences of various combinations of lysine and alanine appear repeatedly in the molecule. Although the elastic element contains some hydroxyproline, this is less plentiful than in collagen.

Covalent bonds cross-link elastin polypeptide chains into a fibre network. After the newly formed elastin polypeptides have been transported out of the cells into the extracellular space, oxidative deamination of certain lysyl residues forms reactive aldehyde derivatives. The desmosine compounds which form the interchain links are derived from four of these derivatives. The derivatives are presumably brought into juxtaposition by folding and alignment of the polypeptides in such a way that condensation results in formation of desmosine (Uitto, 1979).

Elastin appears to be synthesised by connective tissue cells according to the same basic principles as other mammalian proteins. Even though the metabolic turnover of connective tissues and in particular elastin is relatively slow, there is normally continuous degradation of small amounts of elastin, probably by elastase derived from leukocytes or cells of similar origin. So, unless synthesis continues, there will be a decrease in the amount of elastin present. Certain elastases in the body have been shown to increase in activity with advancing age and the change in balance between synthesis and breakdown could account for the loss of elastic response in aged skin.

The correlation of mechanical stress with elastic tissue is significant. It has been identified as stimulating the development of the fibres (Fullmer, 1960) as well as being necessary for the production of elastin during fibrosynthesis (Bhangoo & Church, 1976). The number, diameter, length and organisation of fibres within the tissue are influenced by the way in which mechanical forces are applied to the tissue (Cooper, 1969). It is interesting to speculate on the origin of elastotic degenerative changes in the reticular dermis. The changes appear as clusters of deep-staining, coarse, thick elastic fibres interspersed with numerous granules within the superficial layer (Enna & Dyer, 1979). The presence of large quantities of elastic tissue suggests that it has a role complementary to the functions of collagen, perhaps to absorb the impact of forces and return tissue to its original state of tension. In contrast, it would appear that with age it is the sub-epidermal plexus that shows a marked decrease in elastin content and not the deeper dermal layers (Cowdry & Cowdry, 1950). This, at least, would provide a superficial explanation for the onset of wrinkling in the middle years.

GROUND SUBSTANCE

The third type of extracellular material present in skin is commonly called the ground substance, and contains proteoglycans (glycosaminoglycans). These sugar-containing substances can be readily identified histologically and measured biochemically. There are three major proteoglycans in human dermis: hyaluronic acid, dermatan sulphate and chondroitin sulphate. They are present throughout life although the proportions change with age. The proportion of chondroitin sulphate in early childhood is 8%, decreases at puberty to about 2%, but may be found in higher concentrations in the elderly. The proportions of hyaluronic acid and dermatan sulphate are highest in the late teens (55% and 44% respectively), but do not change by more than a few percentage points throughout life, although the total proteoglycan concentration is lower in the elderly (Gillard *et al.*, 1977; Bouissou et al., 1977).

Measurement of proteoglycans suggests that total concentrations vary according to the loads applied to the site. Determination of glucuronolactone in human dermis gave values of between 1 and 2 μm/mg dry weight of tissue in the non-weight-bearing areas of the sole of foot and fingers of the hand. In the weight-bearing areas, values of 3–5 μm/mg dry weight have been reported for men and 1.5–3 μm/mg dry weight for women. The values for men were always significantly different from those of women of the same age (Gillard *et al.*, 1977).

The function of the proteoglycans is probably related to the provision of normal suppleness and turgor. The small decrease in hyaluronate in old age could be related to changes in the physical properties, but it is the water-retention properties of the proteoglycans that maintain and control the hydroelastic properties of skin. One would expect that a large 'domain' would be controlled by each molecule. A mass of water 15 times that of the proteoglycan can be bound in gel form and thus up to 1000 times the molecular volume can be controlled by these molecules. In the large 'domain' salts move freely, but large molecules such as proteins are excluded by the nature of the highly charged polysaccharide chains. In skin polyglycans, uronic acid and the variable quantities of sulphate in dermatan and chondroitin provide the negative charges for ionic binding to positive ions. However, an ion-exchange function in skin has yet to be demonstrated.

In addition to their role as filters, preventing large molecules from diffusing through their 'domain', there is evidence that negatively charged molecules interact with cells and can in certain circumstances direct their migration, growth and differentiation. Exactly what the proteoglycans do in relation to growth of fibrous connective tissue is not clear. There is a generally held belief that they are intimately related with the collagen fibres, providing them with a coat or outer layer, and that in later life they play an important part in the stabilisation and aggregation of the fibres. Whether the proteoglycan also directs the orientation of fibres in skin remains speculation.

Degradation and turnover of proteoglycans have not been clearly demonstrated in skin, although polysaccharide-degrading enzymes have been found. These enzymes can act upon hyaluronic acid and chondroitin sulphate, but not on heparin sulphate, dermatan sulphate or heparin, and no proteases which act upon the protein core of the proteoglycans have been demonstrated. Thus, it is not yet known whether there is turnover of the entire molecule or only of small portions. Although skin fibroblasts have been shown to degrade polysaccharide, the inability to degrade proteoglycans has been used to diagnose proteoglycan storage diseases (Pinnell *et al.*, 1979). The proteoglycans, therefore, appear to be a relatively stable component of skin, but more knowledge of both structure and function of these very large molecules is required.

DERMAL VASCULATURE

The arteries of skin pass through the deep fascia at definite points and thereafter in some sites branch and anastomose to form the first plexus, which lies in a plane parallel to the body surface and superficial to the deep fascia. From this plexus branches pass to the adipose tissue of the hypodermis where they give rise to the capillary networks between fat cells. Other branches pass through the hypodermis to form a second plexus in the deep layer of the dermis. From this plexus, branches pass to the various structures of the dermis: to form capillary networks through and around the coiled sweat glands, focal accumulations of fat cells and the papillae of the hair follicles. Further branches rise to give a third plexus in the sub-papillary or outer zone of the dermis and to provide arteriovenous anastomoses.

From the sub-papillary plexus, arteriolar branches rise to supply the capillary loops of the dermal papillae. It is these dermal loops growing perpendicularly towards the surface that are commonly observed by capillary microscopy. The number of loops varies from one area of the body to another, being few in cheek but very numerous in fingers, toes, palms and soles. The capillaries proceed to near the tips of the papillae and frequently come into close juxtaposition with the epidermal–dermal junction before they turn abruptly to pass back to the lower edge of the papillary layer of the dermis. Other arteriolar branches supply the sebaceous glands and the remainder of the hair follicle including the arrectores pilorum muscles.

With age, there is a tendency for a reduction in the number of capillary loops present in the papillary region. This is most obvious in areas such as the shin and least obvious in the nail folds (Ryan & Kurban, 1970). Variations in the morphology of the loops are several. In areas of epidermal hyperplasia there is also an accompanying hyperplasia of the blood vessels, which may become increasingly tortuous and coiled. In older skin there is a tendency for the calibre of the vessels to increase, with aneurysmal dilatation of the loops. Variations can occur in 'normal' skin, too, in the level of origin of blood vessels. An arteriole entering the deep dermis from the second plexus may pass without interruption to the horizontal papillary plexus or it may send branches to a hair follicle or sweat gland as it ascends through the corium. There are many interconnections between arterioles and venules at all levels of the dermis, but despite this, the three principal plexuses remain the most important vascular meshes in skin.

The papillary plexus contains most of the dermal blood vessels, and is composed of terminal arterioles, capillaries and post-capillary venules. It is the venules that are most commonly seen. They are relatively uniform in diameter, ranging from 18 to 23 μm in forearm skin (Yen & Braverman, 1976). Arterioles, 17 to 26 μm in diameter, have elastin and smooth muscle in their outer walls, while post-capillary venules of comparable size have only pericytes in their walls. In human vessels an interrupted elastic layer is still present even in 20-μm arterioles; the elastin gradually disappears from the wall to form an *external* sheath just before the beginning of the capillary segment. The diameters of the terminal arterioles vary from 7.5 to 12 μm, those of capillaries from 4 to 6 μm, and of venules from 8 to 26 μm. All these figures are from fixed material inspected by electron microscopy and are probably 30% smaller than in the living state, because of tissue shrinkage.

The presence of multilaminated basement membranes is normal in human skin; they may measure from 25 to 100 nm, and separate the endothelial cells from the pericytes. The striking feature of dermal blood vessels in the papillary layer is the structure of the vascular wall, which is composed of the endothelial cells, a basement membrane in which are embedded elastin, smooth muscle cells or pericytes, and variable amounts of individual collagen fibrils. The wall adjacent to the endothelial cell can be as wide as 5–6 μm. The veil cells characteristically surround the smallest vessels, which together with the components listed above, form a structural and possibly mechanical support for the vessels against shearing forces.

Sphincters are important in the control of blood flow through capillary beds. Rhodin (1973) located and analysed the precapillary sphincter area in the microcirculatory bed of rabbit dermis. Rapid change in the colour of skin can be obtained by closing down these sphincters or by physically collapsing the capillary bed. Anastomoses between the arteriolar and venous plexuses in the papillary dermis allow control of pressure in the vessels and hence in the tissues surrounding them.

Trauma to the skin of a particular intensity results in new vessel formation. These vessels develop in a pattern similar to that already present. Mononuclear cells close to the epidermis have a phagocytic function, and in the control of healing and restitution of a normal epidermal–dermal relationship the removal of cell debris is an important factor. Unless such removal is effected, the permeability and fragility of the microvascular system is increased (Houck, 1968). The pericytes of the vascular system and Langerhans cells of the epidermis can perform these activities. Any failure results in the arrival of macrophages from the systemic circulation. Wyburn-Mason (1958) suggested that if the extracellular debris is not removed effectively, the endothelium is liable to be made more permeable and may proliferate. This proliferation, it has been suggested, is one of the factors in disorders such as vasculitis.

The presence of port wine stains in skin has been a continuing clinical problem. The central abnormality appears to be an increase in the number of vessels and ectasia (dilated vessels). Vessel number is maximum in the papillary region, but vessel area remains fairly constant throughout the dermis (Barsky *et al.*, 1980). The stain, therefore, could represent a progressive ectasia and aneurysmal dilatation of the cutaneous vascular plexus within a local area. If the stain is related to only one plexus then treatment with laser radiation would improve markedly the condition, but if all plexuses are involved then destruction of one plexus may still result in at least the partial return of the condition after vascular regrowth. Experiments with pulsed laser systems destroying vessels at selected depths will provide fascinating information in relation to such disorders.

A major function of cutaneous blood flow is to regulate body temperature. Over most of its great range, skin blood flow far exceeds the metabolic needs of skin. The vessels, both arterioles and venules, are under reflex neural control from a rich sympathetic adrenergic nerve supply, especially in acral regions. These tonically active vasoconstrictor nerve fibres are the efferent arm from four types of sensory input: (*a*) the cutaneous thermoreceptors, (*b*) the baroreceptors of the arteries and cardiac–pulmonary systems, (*c*) chemo-receptors, and (*d*) receptors activated by upright posture and exercise. Human skin is also unique in that it responds to signals originating from central thermoreceptors to control skin blood flow over the whole range and over most of the body.

The veins are richly innervated by sympathetic adrenergic nerves, which reflexly control their volume, but their responsiveness is modified by changes in local temperature and, while the veins are not responsive to baroreceptors, they do constrict during exercise (Rowell, 1977).

Changes in blood flow with age have been noted. ^{133}Xe clearance is frequently used to measure flow through a region of skin. Although measurements from a large number of patients show a high degree of scatter, there is a marked decrease with age. Tsuchida (1979) gave as the clearance rate from the deltoid region:

$$\text{Clearance rate } (Y) = 0.265 - 0.002 \, x \, (\text{age})$$

It is interesting to note that Tsuchida's study confirmed that skin temperature did not measure skin blood flow. Change in temperature is influenced by heat production in the surrounding and underlying tissues. Blood flow in the face and pectoral regions is higher than in the deltoid muscle, and flows in the posterior cervical, lateral thoracic, lateral abdominal and gluteal sites are lower.

Shortly after skin is injured, the area for several centimetres around the site gradually becomes reddened. This flare, which is a component of the inflammatory response, results from neurogenic stimulation. The threshold of pain in the site is reduced. Evidence has been presented by Chapman (1977) which indicates that a vasodilator peptide, similar to a kinin or neurotensin, is responsible for the transient shift in vascular tone, and this provides further support for the theory that hereditary sensory neuropathies are influenced by afferent innervation. Morphologically, however, the relationship between afferent endings and the microvascular bed remains poorly defined. Nevertheless, it is postulated that all neurons in addition to generating and spreading electrical phenomena, have secretory functions, and these secreted substances affect directly the vascular tone of skin.

INNERVATION OF THE SKIN

The varied sensations arising from skin are derived from a diverse population of cutaneous nerve endings or receptors. Thus, tactile, temperature and pain sensations are each subserved by different groups of receptors. Pain results from the activation of nociceptors that are specialised not only as fast ($A\delta$ fibres) and slow (C fibres), but also with regard to the stimulus modality, viz. mechanical damage or heat. Temperature is coded by the activity of specific warm and cold receptors and tactile sensibility is the product of cutaneous mechanoreceptors.

INTRA-EPIDERMAL NEURITES

Despite the demonstration of intra-epidermal nerve fibres in mammals, no generally satisfactory demonstration has been made of their existence in adult man. They are present in foetal skin and in some specialised areas notably in the eyelids, around the genitalia (clitoris and glans penis), in the vestibule of the nose, and are sometimes found in the epidermis of healing wounds.

DERMAL MECHANORECEPTORS

Most of the data on mechanoreceptors derives from animal study where detailed classification of at least 11 types of ending has been proposed (Horch, Tuckett & Burgess, 1977). Their morphological appearance is equally varied and ranges from free simple endings through Lancelot, pallisade or circular endings, to the loosely laminated corpuscles such as Meissner's corpuscles, mucocutaneous end organs, Golgi–Mazzoni organs, Ruffini endings, Merkel cells and the extensively organised Vater–Pacini corpuscle (usually contracted to Pacinian corpuscle) (Fig. 27).

The general body response to tactile stimulation is given in the following list:

(i)   C-type endings respond to slowly moving or lingering stimuli and may give an after-discharge.
(ii) $A\delta$ endings (axon δ fibres) respond equally well to increasing or decreasing intensity of stimulation, either directly to skin, to guard hairs or to down hairs, and are very sensitive to hand tremor.

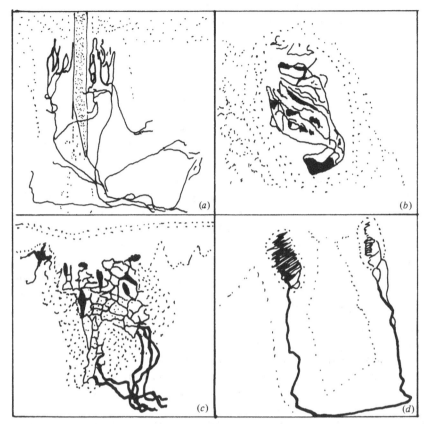

Fig. 25. Nerve endings in skin. (*a*) Petrichial endings. Seen around hair fibres as a varicosed parallel set of nerve fibres in the form of a 'stockade'. (*b*) Mucocutaneous end organ. Sometimes known as Krause bulbs or genital corpuscles. Seen as coil or roll of fine nerve fibres frequently encapsulated. (*c*) Haarschieben. Seen as thickened epidermal discs near or adjacent to hair follicle orifices. Not very prominent in human skin except on the neck and abdomen. (*d*) Meissner corpuscles. Found mainly in dermal papillae as highly organised endings sensitive to pressure.

   (iii)   The field type (guard (G) and field (F) group) of receptors largely associated with hairs responds mainly to streams of impulses.
   (iv)   Ruffini A endings respond to steady deformation maintained for more than 30 seconds and have a resting discharge of about 15 pulses per second.
   (v)   Ruffini B endings may respond to stretch with an evoked response of 15–20 pulses per second, but no resting discharge.
   (vi)   Haarschiebe, a Merkel cell–neurite complex, is a slowly adapting, highly sensitive touch receptor.

Receptors in glabrous skin are equally complex and conform to the following pattern:

   (i)   Aδ receptors are very sensitive to both indentation and retraction changes.
   (ii)   Using the Aα axons, Pacinian corpuscles are extremely sensitive to vibration,

responding to stimuli up to 500 Hz even when of small amplitude. Because most Pacinian corpuscles lie in the deep dermis or hypodermis, they do not appear to have a localised receptive field and do not have a plateau response (constant rate).

(iii) Both Merkel and Ruffini endings appear to respond to maintained steady indentations persisting for more than 30 seconds.

(iv) Some field receptors, distinct from Merkel and Ruffini endings, respond to slow inwards movement of the skin and continue to respond for several seconds; others respond to faster movement, but these have no plateau response; a third type appears to be intermediate between these two, being more sensitive to indentation than retraction and having no plateau response in the absence of vibration.

In physiological testing systems, mechanoreceptive units are usually classified as slowly adapting or rapidly adapting. In human hand (glabrous skin), two types of each have been identified of which the second slowly adapting unit is not commonly found in non-human primates. Recordings from the median nerve about 10 cm from the elbow can be displayed on an oscilloscope. Mechanical stimulation from small blunt probes permits the mapping of receptive fields. Using von Frey hairs with forces ranging from 0.1 to 100 mN, Johansson (1978) obtained general receptive field areas. From skin indentations of controlled amplitude, velocity and position, more detailed mappings were then plotted. Fig. 26, adapted from Johansson's data, shows the distribution of some of the slow and fast receptive fields with their relative areas. As expected the receptive field area increases with indentation amplitude. Measurements up to 25 times the threshold amplitude indicated that field sizes up to 30 mm^2 were obtained with the first type rapidly adapting receptor (RA) and the first type slowly adapting receptor (SA1), but that both of the other types of receptors (PC and SA2) had field sizes up to 80 mm^2 with only 5 times the threshold indentation amplitude.

Responsiveness to stretch across the sensitive zones differs at the same sensitive point depending upon the position of the hand – extended or relaxed. Stretch in the direction of the arrows in Fig. 26 gave the most pronounced discharge, whereas stretch in the direction of the broken arrows decreased the discharge. From these and many other studies it is clear that the mechanoreceptors in skin have the capacity of encoding mechanical events restricted to the skin surface with a high degree of spatial selectivity. The end organs of the nerves are located close enough together to provide a sharply defined area with practically uniform sensitivity. There are enough endings in glabrous skin scattered over the whole area to enable interdigitation and overlap so that analysis of spatial aspects of events at the skin surface gives good resolution. Such a system appears to be well adapted to the analysis of intensity as well as the detection of a number of small simultaneous stimuli. Direction and magnitude and rate of change of tension can also be detected, with the deeper units acting as the sources for large force reception.

Rapidly adapting mechanoreceptors (end organs) are innervated by axons

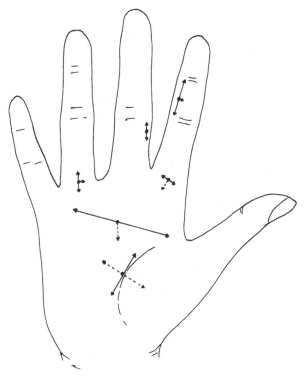

Fig. 26. Sensitivity to direction in stretch zones in skin of the hand. Greatest discharge was effected when stretched in direction of arrows. Broken lines indicate decrease in activity. (After Johansson, 1978.)

whose collaterals in the spinal cord have a distinct morphology, different from those of the Pacinian corpuscle, and also from other axons innervating other cutaneous and muscle receptors (Brown, Fyffe & Noble, 1980).

VATER–PACINI CORPUSCLES

Vater–Pacini corpuscles are highly ordered structures, the largest found in skin, that lie in the deep dermis, along fascial and connective tissue planes, and near nerve trunks. Depending on location, they vary from 0.2 to 1.0 mm. They are most numerous in the digits, the penis and clitoris. Each corpuscle has a stalk in which the receptor, a heavily myelinated nerve, makes several turns before entering an oval, concentrically laminated capsule (Fig. 27). The concentric fibrous laminae have fluid spaces between them. The spaces between the laminae increase from the innermost towards the periphery of the corpuscle.

When a displacement of the outer laminae is caused by an appropriate pressure, the stimulus is transmitted to the core, producing excitation in the

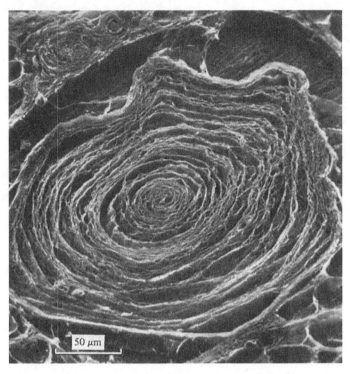

Fig. 27. Scanning electron micrograph of a section through a Pacinian corpuscle.

neural terminal and giving rise to the receptor potential. Observed on an oscilloscope, spike potentials of about 70 μV can be elicited during the early or late depolarising phase. Increasing stimulus intensity lengthens the time of peak signal. Maximum response frequency is at about 500 Hz, but the ending will respond to stimulus frequencies from about 100 to 3000 Hz. Above a stimulus frequency of 500 Hz the peak response potential remains constant.

A mathematical model has been proposed by Grandori & Pedotti (1980) defining the transfer function (G_1) for a given set of conditions in terms of the static gain (u_1) and appropriate time constants T_x for the system, where:

$$(G_1) = \frac{u_1\,(1+T_1)}{(1+T_2)\,(1+T_3)}$$

The mechano to neural transduction was modelled as:

$$(G_2) = \frac{u_2}{(1+T_c)}$$

where T_c is the time constant derived for the particular set of conditions at the boundary between corpuscle lamina and nerve ending. The values of time constants, T_1, T_2 and T_3, were obtained by using a fitting procedure on the experimental data. The value of u_1 was taken as 1.0×10^{-2} and u_2 as 1.0. The proposed model interprets receptor potential time-course and stimulus shape for both intact and decapsulated corpuscles. One aspect is worth noting because it provides a possible explanation for the second depolarising response, or off-response, of the corpuscle: the model suggests that this second response is the result of concerted effort from the dynamic action of the lamellated structure.

Since the corpuscles are known to respond to vibratory pressure and not to steady pressure, their site near blood vessels has led to the suggestion that they are probably concerned with local changes in blood supply and control of arteriovenous anastomoses (Sinclair, 1967). Individual collaterals from axons of the Pacinian corpuscle carry up to about 370 synaptic boutons at the spinal cord, indicating a high level of signal distribution (Brown *et al.*, 1980).

MUCOCUTANEOUS END ORGANS

Found underneath glabrous cutaneous epidermis and near mucocutaneous surfaces, end organs are structurally organised coils or rolls of fine nerve fibres, sometimes encapsulated, from which issue myelinated A fibres. Winkleman (1960), who named these structures, suggested that without the presence of the hair follicle, the nerve net seen around the fibre would curl up into a ball and move closer to the surface to appear as a mucocutaneous end organ. This seems to happen in the labia minora when the hair follicles and associated nerves of the human female foetus and infant disappear to be replaced by the 'genital corpuscle' or mucocutaneous end organs.

Structurally, by comparison with Meissner and Pacinian corpuscles, end organs seem to have an intermediate degree of organisation. They comprise loops of loosely wound, branching, non-myelinated nerve fibres forming an irregular oval mass surrounded by a condensation of connective tissue. They have also been described as Krause end bulbs, genital end bulbs and Dogiel's body. The corpuscle of the end organ is divided into sub-lobular units of axon terminal surrounded by concentric lamellar cell processes (MacDonald & Schmitt, 1979). The interlamellar substance contains elastin and collagen including the coarse-banded type. Specialised zones of contact connect the lamellar processes and axons. The similarity at the ultrastructural level to Meissner corpuscles suggests that they are merely an organisational variation. Specifically, they are to be found in the sub-papillary dermis of the clitoris, prepuce, glans penis, lip, tongue, gum, eyelid and perianal area, but they do *not* appear in true mucous membranes or hair-bearing skin.

MEISSNER CORPUSCLES

This nerve ending is found consistently in dermal papillae, frequently near the summit to lie close to the epidermis of volar skin of hands and feet. The corpuscle is often single, about 100 μm long, and is outlined by a definite but thin capsule of connective tissue. Within the capsule a series of elongated conical cells with their bases set on the capsule and their apices almost touching at the opposite side give the corpuscle, in longitudinal section, a layered appearance (Fig. 25 *d*). A large-diameter myelinated A fibre divides into non-myelinated branches before passing into the organ where the branches form fine neurofibrillar expansions over and between the conical cells. The nerve fibres stain readily for non-specific cholinesterase.

RUFFINI ENDINGS

Ruffini endings are found in many parts of the body, and can be seen after Methylene Blue staining as spray endings lying on or between bundles of collagen fibres. They can also be found in the walls of certain blood vessels. The myelinated nerve fibre branches several times to form a number of fine non-myelinated fibrils at the end of which lie densely staining knobs. The tendon organ (organ of Golgi) is composed of a number of such units. Ruffini endings occur in skin as discrete organs and may be larger in the deeper zone of the dermis.

HAARSCHIEBEN

Thickened epidermal discs adjacent to the hair follicle orifices, the *Haarschieben* of Pinkus, are special touch receptors in the epidermis. They are served by numerous nerves from the neck of the follicles, are 100–300 μm in diameter and are present at a density of about 1 per square centimetre. The thickening of the epidermis is the result of two additional layers of pseudostratified columnar cells filled with fibrils and orientated perpendicular to the skin surface that extend down to the basement membrane. The highly complex epidermal–dermal junction contrasts with the normally smooth junction of surrounding tissue. Prominent rete pegs often mark the presence of *Haarschieben*. The special basal cells, Merkel cells, form a complex with neurites which in animals is a highly sensitive, slowly adapting, modality-specific touch receptor. However, studies on human skin reveal equivocal responses from these structures.

In man, the *Haarschieben* vary widely in number from subject to subject and are sometimes difficult if not impossible to identify under the dissecting microscope (Smith, 1977). They are more abundant on the neck and abdomen. The Merkel cell which forms part of the complex is a distinctive cutaneous cell migrating from the neural crest to the skin. Histochemically it is similar to other neural crest end organs, possessing similar enzymes

(Winkleman, 1977). It is similar to the neuroepithelial cells of the taste bud and is believed to release a polypeptide hormone. Many other cells of neural crest origin throughout the body also produce or are believed to produce similar polypeptides. The granules within the Merkel cell are thought to be associated with this peptide production.

Many large capillaries are also seen in association with the *Haarschieben* and it is the presence of these vessels that gives the skin its characteristic colour or spot. There are many half-desmosomes between the basal cells and the basement membrane, but there seems to be no synaptic speciality between the Merkel cells and the neurites, although it is generally thought that some specialisation exists. Sometimes the Merkel cell granules have been seen to fuse with the specialised area of membranes and this has been taken to indicate synaptic transmission.

In animals, the *Haarschieben* transmit neural information to the spinal cord proportional to both rate and degree of indentation of the skin surface. The frequency of discharge becomes progressively greater during the initial phase of indentation and can achieve 1500 pulses per second. The frequency is related to the velocity and amplitude of the displacement. In man, the *Haarschieben* is not a dead spot, but seems to have little more sensitivity to touch, pain and tickle than the surrounding skin. It is not exquisitely sensitive, as it is frequently found to be in animals.

PETRICHIAL ENDINGS

There are a number of different nerve fibres associated with the hair follicle. One of the most intriguing lies at the level of the bulb of small or vellus follicles and at the bulge of large follicles. It is seen as a basket of varicosed parallel fibres in the form of a 'stockade' (Fig. 25a) (Heyden, 1969). Very fine non-myelinated fibres pass into the papilla of the root bulb. These are part of the ground plexus and are probably not associated with sensory functions. Fibres from these ground plexuses also pass to the arrectores pilorum muscles where they have motor function. Many fibres appear to end as 'free endings' both in association with the external root sheath, where palisades of endings can be seen, and in the papillary region of the dermis. However, by electron microscopy, true free endings have never been seen. Each ending has either a specialised terminal structure or is associated with a cell, usually a specialised cell (Breathnach, 1977).

SENSATION OF PAIN

There is still some confusion over which endings can and do give rise to the sensation of pain. Chemical mediation of cutaneous pain both direct and referred can be attributed to histamine. Repeated injections of histamine liberators continue to cause pain, but the associated responses – itch, flare and weal – diminish. Rosenthal (1977) believes that histamine released locally in the epidermis is the chemical mediator of pain, the receptors being the epidermal nerve endings. Other types of pain are thought to be related to intradermal pressure or pressure in the sub-dermal tissues, with the various mechanoreceptors being involved in transmitting the sensation.

CHEMICAL MEDIATORS IN SKIN

The discovery that central and peripheral neurons contain a wide variety of peptides, many of which are either the same as or similar to the peptide hormones released by endocrine cells and specialised cells in tissues, has dramatically changed thinking about regulatory systems. Nervous and endocrine systems have been regarded as separate control mechanisms, but are now seen to share a set of chemical messengers that are delivered to target organs or cells by a variety of modes. These include three pathways all present in skin: the endocrine (cell to distant target), the paracrine (cell to neighbouring target through the interstitial spaces) and the neurocrine (nerve to target across a narrow synaptic gap). The paracrine system may also be involved in sensory mechanisms. This exciting unifying hypothesis has led to the search for similar control systems in other parts of the body, for example in the supply of messengers to the digestive system (Grossman, Brazier & Lechago, 1981). The same chemical messengers may be used for fast as well as slow control of function, the mode of delivery and site of action determining the overall effect. At the present time the answers to many of the questions this hypothesis presents are only just beginning to emerge, but the inter-relationship of nervous control, through adrenergic and cholinergic endings; sensation, through the small peptides released by Merkel and other cells; rate of cell division, through the chalone system; and the many other functions of cells such as the Langerhans cell, offers a burgeoning, exciting and promising field of study.

MORPHOLOGICAL DIFFERENCES IN SKIN AREAS OF THE BODY

HEAD

Scalp All layers including sub-cutaneous tissue (with fat and occipito-frontalis muscle or its aponeurosis) bound together tightly.
  Thin epidermis; irregular dermal papillae; elastic fibres numerous; hair follicles long; large sebaceous glands; prominent arrectores pilorum muscles; sweat glands.

Forehead Fine downy hair; generally no dermal papillae; sweat glands numerous; large sebaceous glands.

Eyebrows Long hairs without arrectores pilorum muscles; large numbers of thick elastic fibres.

Eyelids Thin epidermis and dermis; few fine hairs with sebaceous glands;

few sweat glands; no fat in dermis or in underlying tissue; some fine elastic fibres; some cells in dermis contain pigment.

Face　Thin epidermis; small dermal papillae; striated muscle fibres penetrate into dermis; focal and twisted groups of elastic fibres; melanoblasts; large sebaceous glands.

Nose　Large sebaceous glands.

Lips　Margins sharply defined by colour change; epithelium thick but very transparent; long conical papillae; deeper dermis contains groups of fat cells.

　Oral aspect: shorter dermal papillae; salivary glands; striated muscle of orbicularis oris.

Pinna　Skin closely bound to cartilage; dermal papillae absent or few; arteriovenous shunts or anastomoses prominent.

BODY

Axilla　Thin epidermis; long hairs in Caucasian adults of both sexes; eccrine sweat glands; apocrine glands; sebaceous glands of hairs.

Areola of breast　Very pigmented; elastic fibres numerous, but less than nipple; eccrine sweat glands; apocrine glands; smooth muscle bundles both circular and radial extending into breast skin.

Nipple　Moderately thick epidermis; dermal papillae compound; no sweat glands; sebaceous glands *without* hairs; smooth muscle bands; thick elastic fibres; elastic arches in dermal papillae; various groups of nerve endings.

Abdomen　Thin epidermis; elastic fibres short, parallel to surface; few eccrine sweat glands; regions of fat in dermis; very thick fatty hypodermis.

Back　Thin epidermis; thick dermis; elastic fibres numerous and thick; few eccrine sweat glands.

Labia majora　Hairs on outer surface; inner surface without hairs; thin layer of smooth muscle; eccrine, apocrine and sebaceous glands.

Labia minora　Epidermis pigmented; long dermal papillae; no fat cells; fine elastic fibres; large sebaceous glands *without* hairs; mucocutaneous end organs.

Penis　Thin, pigmented skin loosely attached; no fat in dermis or hypodermis; large sebaceous glands; Meissner and Pacinian corpuscles.

　Glans: no sweat glands.

　Praeputeal fold: modified sebaceous glands.

Scrotum　Thin, pigmented epidermis; sebaceous glands; *no* elastic fibres; smooth muscle deep.

Perineum　(including circumanal region) Thin, pigmented skin; sweat glands; apocrine glands; smooth muscle bands; thick elastic fibres; no dermal papillae in mid-line.

　Male: hairs in peri-anal region.

LIMBS

Arms   Thin epidermis; elastic fibres numerous and thin; eccrine sweat glands; fat in deeper dermis; thick hairs on surfaces; sebaceous glands.

Finger pads and toes   Glabrous; very thick epidermis firmly bound to tissues below; epidermal ridges; deep stratum granulosum and well-marked stratum lucidum; dermal ridges and compound papillae; numerous sweat glands; *no* sebaceous glands; arteriovenous anastomoses.

Palms   Glabrous, thick epidermis; epidermal ridges; stratum granulosum and stratum corneum well marked; dermal papillae very long and compound; few thick elastic fibres; sweat glands numerous; *no* sebaceous glands; arteriovenous anastomoses; Pacinian corpuscles.

Sole of foot   Glabrous, thickest epidermis; stratum granulosum and stratum lucidum well marked; epidermal ridges; long compound dermal papillae; sweat glands numerous, but deep in dermis; *no* sebaceous glands; arteriovenous anastomoses; Pacinian corpuscles.

Nail bed   Marked dermal ridges; arteriovenous anastomoses; *no* sweat glands or sebaceous glands.

4

MECHANICAL PROPERTIES OF SKIN

Skin acts as a mechanical barrier between the interior of the body and the outside world, but is required to glide and stretch during body movement. Thus, resistance to mechanical force is a necessary property and if this is reduced, as in epidermolysis bullosa, stripping of the epidermis by relatively low shearing forces may cause life-threatening complications. On the other hand, scleroderma (thickening and tightening of the skin) may, in the later stages, inhibit body movement. Surgery, especially plastic surgery, is much concerned with skin extensibility and surgeons develop a sensitivity to and an awareness of the quality of skin from different parts of the body.

Before we can discuss the mechanical properties of skin, some definitions of common terms are necessary, since these are often misunderstood or misused. *Load* is the total force applied to the specimen but *stress* is the force per unit area where area is measured at right angles to the line of action of the force. Similarly, *extension* is the increase in length in a tensile test while *strain* is the change in length per unit length. Strain may also be expressed as a percentage. If skin is removed from the body, retraction occurs indicating that skin normally exists in a state of tension. The amount of retraction varies with site and direction (Langer, 1862*a*) and is an indication of *anisotropy* or variation of properties with direction. Retraction does not occur instantaneously and so skin is said to be *visco-elastic*. The mechanical properties of visco-elastic materials alter with the rate of loading or straining. For skin, stress, strain and time are not related by simple linear equations of the type $y = ax + b$ and so skin is an anisotropic, non-linear, visco-elastic, inhomogeneous material.

TENSILE TESTS *IN VITRO*

For tensile tests *in vitro* an excised specimen of skin is gripped in the jaws of a suitable device. The length and cross-sectional area of the specimen are measured and either the load is applied at a set rate of increase and the extension measured or the jaws separated at a pre-determined rate and the load measured. Stress and strain are then calculated. However, before attempting any mechanical test it is important to control the pre-test and test conditions and the following points must be considered:

1. Skin removed from the body retracts; thus, either the original specimen dimensions should be measured on skin whose configuration closely corre-

sponds to that *in vivo*, remembering that even the way of laying out a body affects the pre-existing skin tension (Langer, 1862*a*), or specimen dimensions can be taken from skin which has freely and completely retracted. The conditions under which measurements are taken should always be stated.

2. Skin is a living organ and after death changes will occur. The storage of skin between death and excision and testing may affect the mechanical properties. After excision, short-term storage is best carried out in a sealed container at 4 °C and experience indicates that overnight storage is acceptable. For longer periods, tissue is best handled by freezing in sealed containers. Tests seem to indicate a greater scatter of data after freezing. In experiments carried out in our own laboratories, the mean did not shift significantly, although the numbers of specimens were not sufficient for full statistical analysis (Al-Haboubi, 1977).

3. Skin has a 'memory' for previous mechanical loading and so the forces involved in excising the skin and removing the sub-cutaneous fat should be as small as possible.

4. When left in the normal room atmosphere, skin will tend to dehydrate. Jansen & Rottier (1958) showed that desiccated skin is stiffer than normal. During preparation, skin may be kept in damp tissues or blotting paper, but precautions to prevent drying must be taken during testing. Immersion in a fluid prevents dehydration but permits imbibition of fluid (Viidik, 1973). In addition, buoyancy effects decrease internal loading due to reduction in effective weight of the specimen. Spraying with a fine mist of water is probably adequate for fairly rapid testing, but immersion or a high-humidity chamber is required for longer test times.

5. Gripping the specimen may cause problems. The specimen must not move within the grips, but if the grips themselves produce stress concentrations then the specimen may fracture outside the volume under test. The use of adhesive to attach the specimen to subsidiary grips, which are then held in the machine clamps, may be advantageous (Al-Haboubi, 1977).

6. Devices for measuring load and extension must be appropriate to the test regime, remembering that skin extensions are typically 50–100% before rupture. Available devices have been reviewed by Stevens & Jones (1977).

The calculated stress–strain curve from a typical tensile test is shown in Fig. 28 and indicates that as length increases, cross-sectional area must decrease (Kenedi *et al.*, 1975). If linear strain is plotted against volume strain then at higher linear strain the volume decreases, possibly due to the expulsion of fluid (Gibson, Kenedi & Craik, 1965). Because of the complexity of the mechanical response it is convenient to discuss the behaviour of skin in three phases: the first, Phase I, representing the rapid extension of skin under low load; the second, Phase II, representing the stiffening of skin; the third, Phase III, representing the final relatively stiff behaviour.

Explanations for the shape of the curve have been sought through analysis of skin morphology, particularly the arrangement of fibres in the dermis.

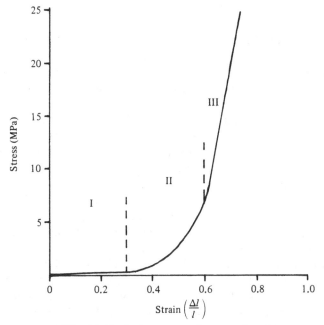

Fig. 28. Typical stress–strain curve for skin.

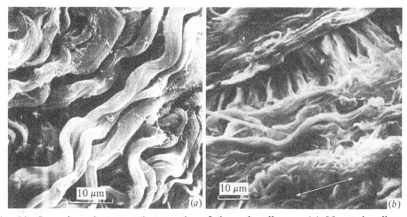

Fig. 29. Scanning electron micrographs of dermal collagen. (*a*) Normal collagen bundles in young skin. (*b*) After extension in the direction of arrows, most of the bundles of collagen have been straightened and tend to lie either in the direction of load or approximately at right angles.

Samples of skin have been extended into the different regions of the stress–strain curve, chemically fixed to preserve the structure, and mounted for examination by scanning electron microscopy. Typical micrographs are shown in Fig. 29. It has been clearly demonstrated that as load is applied to the tissue, the fibres begin to orientate in the direction of stress and to straighten (Brown, 1973). By the end of Phase II most of the fibres are orientated towards the direction of stress and are straight. The load response characteristics of the fibres are determined in Phase III.

If skin is loaded to a point less than the breaking stress and the load removed, the loading and unloading curves are non-coincident; skin exhibits *hysteresis*. If the load cycle is repeated the second curve is displaced and entry to Phase II occurs at higher strains. However, after about six cycles the curves become effectively coincident. Advantage has been taken of this phenomenon to reduce artefacts related to loads applied during specimen handling (the process is termed pre-conditioning), but the results so obtained may mask real differences between specimens.

One of the difficulties in characterising skin is that there are few simple numerical values which can be derived from the stress–strain curves which are well defined. None of the phases is strictly linear, although Phase III is often treated as such. When this assumption is made, the slope of the curve in Phase III is often quoted as the 'Young's' or elastic modulus, the names deriving from an analogy with the linear stress–strain relationship found in metals. The slope of Phase III increases as the logarithm of strain rate (Vogel, 1972). Furthermore, the breaking strength of skin is also related to strain rate. Thus, skin appears to resist fracture at very high strain rates.

In order to obtain some means for comparison of characteristics of specimens tested in different laboratories, it has been suggested that the strain at Phase II be quoted (Kenedi *et al.*, 1975). This value has been given by Millington *et al.*, (1971), Yamada (1970) and Lanir & Fung (1974). It has also been estimated from theoretical arguments by Markenscoff & Yannas (1979). Typical values for young skin are from 0.4 to 0.65.

When skin is cut perpendicular to the body axis the curve shown in Fig. 28 is not always produced. Vogel & Hilgner (1979) reported that at certain loads the extension suddenly increased without a corresponding increase in applied load. This phenomenon has been observed in rat, guinea pig, rabbit and dog skin, but so far not in specimens taken from mouse or human skin. It appears most frequently at a rate of 5 mm/min, diminishing as the strain rate is increased.

It would seem then that there is a paucity of numerical values to characterise the mechanical properties of skin. So far we have values for breaking strength, strain or extension ratio (final length/initial length) at rupture, a rather ill-defined modulus and an equally ill-defined strain at entry to Phase II. Some attempts to define empirical equations which match the observed stress–strain curves have also been made, among them that of Ridge & Wright (1964). Thus many authors resort to indicating the general directions of shift in the curves produced when different factors are varied (Daly & Odland, 1979), which does not allow comparison between results obtained by different laboratories.

TENSILE TESTS *IN VIVO*

Uniaxial tests *in vivo* cannot be compared directly with results from tests *in vitro* because the lateral contraction of the skin is inhibited by surrounding tissue and also because results may be influenced by tethering to sub-cutaneous structures. Testing *in vivo* allows control of the environment and 'state' of the tissue at the expense of defined mechanical conditions. Again some of the factors affecting tests *in vivo* will be considered before evaluating the general results.

1. In order to load tissue some sort of device must be attached to the skin. The outer layers of dead epidermal cells are loose and constantly being shed. If adhesives are used to attach the device these layers should be removed by several applications of adhesive tape (e.g. Sellotape). Also, it is usually desirable to remove the hair by careful shaving. Special care is needed if using cyanoacrylate adhesives since a general tissue reaction may occur on abraded skin.

2. Extension of the dermis must be achieved by loads applied to the surface. These loads are limited by the epidermal–adhesive bond and the epidermal–dermal bond. Extensions of the dermis are inferred from movements of the attachments and so slip or creep in either junction will result in inaccuracies in the measured dermal extension. Attempts to overcome this difficulty were made by designing suction grips for tests *in vivo* (Kenedi, Gibson & Daly, 1965). These have limited applicability because of size and the requirement for relatively loose skin around the test site.

3. Tests *in vivo* should not cause permanent damage. Excessive loads should not be applied, especially to abnormal skin. Reported discomfort should cause the test to be terminated irrespective of the mechanical behaviour being observed. Most subjects, however, can be tested to the start of Phase III without discomfort.

4. Body position affects skin tension and should be standardised within each experimental series and be specified in all reports. Effects are most noticeable near joints and in areas where there is little redundant skin.

In general terms, the shape of the curve characterising a test *in vivo* is similar to that from a test *in vitro*, with the understandable difference that the point of rupture is not reached. In many subjects data can be obtained well into mid-Phase III. During observation of the skin under test it will be noticed that around Phase II or early Phase III the skin loses colour (blanching). The cause is occlusion of the dermal capillaries by the resultant pressure of surrounding tissue. Quite shortly after blanching is observed, the subject usually experiences sufficient discomfort to ask for the test to be terminated. In surgery, if a state of stress sufficient to cause blanching is maintained over a long period of time, then the outcome is usually skin breakdown.

Before any attempt at converting load to stress can be made, the thickness of skin must be estimated. Skin folds usually contain sub-cutaneous fat and using a fold from a relatively fat-free area such as the dorsum of the hand to estimate thickness of skin from other sites involves a lot of assumptions. One of the radiographic (Black, 1969) or ultrasonic (Gunner, *et al.*, 1979) methods can be used on the test site.

Repeated tests on the same area of skin produce a 'pre-conditioning' response similar to that shown in tests *in vitro*, but recovery will occur. The time between tests on the same site should be specified and trends in the values obtained from repetitive tests sought. These recovery times will always be measured in minutes, but vary with the test protocol.

DIAPHRAGM TESTS

A popular method for the mechanical testing of rubbers is to clamp a sheet of rubber over an opening and then to inflate the sheet into a dome, measuring either pressure and volume or pressure and central displacement. Comparable systems have been used to test skin both *in vivo* and *in vitro*. Similar difficulties of specimen collection, storage and preparation are encountered in this form of testing as in the uniaxial tensile test. One advantage of skin over other tissues is that the stratum corneum need not be kept wet and so pressure may be exerted through physiological saline on the dermal side of the specimen, leaving the stratum corneum in air.

Slippage of skin in the clamp and tearing at the clamp edges may be problems if high stresses are used. Spiked clamps (Dick, 1951) or adhesive (Alexander & Cook, 1977) have been used to avoid slippage.

The relationship of volume or displacement change to strain within the area under test rests on certain assumptions. If, as is often the case, central displacement is measured then the geometrical shape of the dome must be known or assumed. Grahame (1970) assumed that with a circular aperture this shape is a spherical cap, in which case strain energy is uniform throughout the specimen and so may be calculated. This assumption held when the apparatus was tested with a sheet of rubber, but the effects of anisotropy were not considered in detail. A rectangular cup with rounded ends having a width to length ratio of 1:4 was used by Cook, Alexander & Cohen (1977). The dome shape was assumed to be an 'inflated cylinder' and this was found to be correct when the apparatus was tested *in vitro* and the dome filled with casting material which could then be sectioned. With both sets of apparatus suction was applied to the stratum corneum *in vivo* and the central displacement measured.

Alexander & Cook (1977) consider skin to be isotropic and that the observed anisotropy is due to pre-tension existing in living skin. A 'pre-tensioning' device was constructed by which the effects of this tension were to be removed. In essence, the skin between two tabs of greater separation than the diameter of the suction cup was compressed in the plane of the skin until wrinkling occurred; this position was retained for 5 minutes, the skin released and the tabs rotated through 45°. The procedure was repeated until all directions had been treated. After this procedure 18 out of 23 patients were said to show isotropic behaviour of the skin on the upper back, a normally anisotropic area.

If the skin will slide freely into the dome then suction techniques may be used to investigate skin anisotropy (Brereton, 1974). Before the test the outer diameter of the dome is marked on the skin, suction applied until the skin fills the dome, and the outer diameter re-marked. It is assumed that the stiffer the skin in a particular direction the more skin will be pulled into the dome, and so lines drawn along the maximum diameter should correspond to the lines of maximum stiffness and maximum pre-tension. If a plot is made of outer diameter versus dome height, a three-phase curve is produced; the *horizontal* Phase III is due to testing the skin within the dome rather than pulling in skin from outside. Absolute values for the mechanical properties of skin cannot be calculated using this apparatus, but changes with direction, site and subject may be demonstrated. One of the difficulties is that results are affected not only by body position but also by the friction between skin and dome.

One of the objections to diaphragm testing *in vivo* is the possible effect of differences in the bond strength between dermis and sub-cutaneous tissue on the results. Studies have shown that there is a negligible vacuum beneath the dome of skin in dogs and that tests *in vivo* and *in vitro* on the same piece of rat skin gave reproducible results (Cook *et al.*, 1977). In practice, if specialised areas of skin such as palm are excluded, variations in bond strength are unlikely to be a major source of error.

VOLUME COMPRESSIBILITY

Skin and other soft tissues might be expected to show coefficients of volume compressibility close to that of water. North & Gibson (1978) used a hydraulic cylinder (Schrader type 491 B) in which a specimen of skin immersed in fluid could be compressed. After calibrating the apparatus with steel and measuring the compressibility of water, glycerine and olive oil, the volume compressibility of skin immersed in these fluids was measured. A value of 0.30 m²/GN was obtained for skin (compared with 0.28 m²/GN for glycerine and 0.42 m²/GN for water). Thus under these conditions skin is virtually incompressible. Such a result could have been predicted from the values of Poisson's ratio, shear modulus and modulus of elasticity (Vossoughi & Vaishnav, 1979). Similar results were reported by von Gierke (1962) but no details of the method were given.

The relative incompressibility of skin does not invalidate observed volume changes in tensile testing *in vitro*, since loss of volume would be due to expulsion of interstitial fluid.

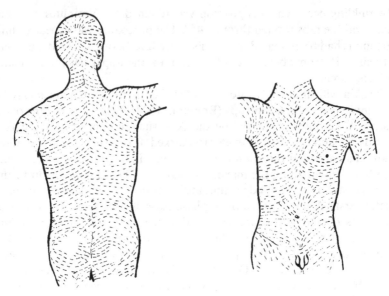

Fig. 30. Langer lines from an adult cadaver (left) and from the cadaver of a 2-year-old boy (right). (For details of other regions of the body see Langer, 1861, 1862*a*, *b*.)

DIRECTION AND SITE VARIATIONS

Early tests on variation of mechanical properties with site and direction were made by Dupuytren (1836) and Langer (1861, 1862*a*, *b*), spurred on by observations on gaping wounds. The well-known 'Langer lines' were demonstrated by puncturing the skin of corpses with a sharp circular awl and noting the long axes of the resulting elliptical holes (Fig. 30) (Langer, 1861). Langer (1862*a*) himself reported that these lines could be affected by methods of laying out the corpse and suspected that they depended both on pre-tension and possibly on mechanical properties. Circular incisions confirmed the effects of pre-tension since circular incisions in an area showing elliptical holes produced a central ellipse of tissue with its long axis at right angles to the elliptical hole in the surrounding tissue. The effect of circular punctures was confirmed by Cox (1941).

Langer (1862*b*) also attempted to investigate the mechanical properties of strips of skin tested parallel to and perpendicular to his lines by hanging successively greater weights from them and noting the extensions. Langer performed his experiments in a glass tube in which moist paper had been placed to increase humidity and was able to show that strips cut parallel to were less extensible than those cut perpendicular to his lines. He also demonstrated the typical load–extension curve found for skin. Major improvements in mechanical testing were not made for a further 100 years.

If the extensions for a given load are measured at angles of 45° around a given site and the results displayed as a polar plot, then the lines joining the minimum diameters closely follow Langer's lines (Stark, 1977).

The origins of this anisotropy may result from two variables which are not completely independent: structure and pre-tension. The early part of the tensile test curve relates to the arrangement and interaction of the tissue components while Phase III reflects the mechanical properties of the collagen fibres themselves. Mechanical forces (extrinsic or intrinsic) tend to affect the arrangement of tissue components and so 'pre-tension' may influence the arrangement of fibres and the response to extrinsic force. For example, during late pregnancy the abdominal skin is stretched to accommodate the growing foetus and this alters the cleavage pattern both during and after pregnancy. Similarly, the limb cleavage patterns of a baby are predominantly circumferential at birth and slowly change to the adult pattern by the age of 2 as limb movements become more energetic.

Skin extension varies from site to site. For example, extensibility on the medial side of the thigh is relatively large to accommodate abduction of the hip while over the ventral aspect extensions are much less. Similarly on the dorsal aspect extensions are large but highly directional just below the buttock to allow hip flexion and fall at mid-thigh. This further emphasises the importance of body position when interpreting the results of mechanical tests *in vivo*.

VARIATIONS WITH AGE

Most systematic studies of variations in mechanical properties with age have been carried out on rats. Although Vogel examined the breaking strength of rat skin after drug treatments, his data for the control groups have been added to those reported in other studies on age relationships and the cumulated results plotted in Fig. 31. Peak strength appears in middle age with a slow decline to old age. Interpretation needs some care as skin thickness increases throughout life in rats, but even taking this into account skin strength still decreases with age. The elastic modulus (calculated from Phase III) also shows a slight drop in old age (Vogel, 1976a).

Explanations for age-related changes have been sought in collagen biochemistry. Four fractions of collagen are distinguished by solubility: Fraction I or soluble collagen dissolves in 0.15 M sodium chloride, Fraction II in 0.5 M sodium chloride, and Fraction III in citrate buffer; Fraction IV or insoluble collagen is the remainder. In rats, Fraction I decreases until adolescence and remains fairly constant thereafter; Fractions II and III diminish sharply during the first 4 months and thereafter fall more slowly; Fraction IV (believed to represent the most highly cross-linked and hence stiffest collagen) increases sharply in the first 4 months and then very slowly until death (Vogel, 1976b). If the 'surface mechanical resistance' (tensile strength in kg/mm²

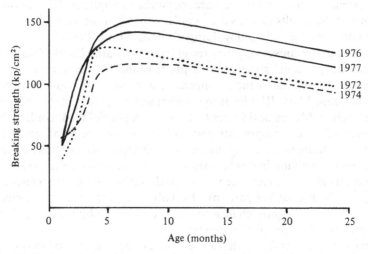

Fig. 31. Variation in breaking strength of rat skin with age. (Data plotted from Vogel, 1972, 1974, 1976*a*, 1977.)

collagen multiplied by collagen per unit area) is used as a measure then skin increases in strength throughout life, an observation consistent with the increase in Fraction IV collagen (Fry, Harkness & Harkness, 1964).

Studies on human tissue indicate a trend for Phase II to occur at lower extensions as age increases. Using the suction cup, the elastic modulus appeared to increase with age and the modulus for females was, on average, higher than that for males of equivalent age. Skin thickness decreased with age, but differences between the sexes are small. The results from children (3–17 years) do not seem to lie on the same regression line as those from the older subjects (Grahame, 1970).

RADIATION EFFECTS

Measurement of radiation effects on skin have been made using a portable hand-held extensometer (Burlin, Hutton & Ranu, 1977). The modulus was measured at a strain of 30% and a steady decrease during therapy and an increase after therapy was noted. A criticism of the procedure adopted was that in order to compute stress, for which cross-sectional area is required, the length of tabs attached to the skin and a measure of skin thickness derived from skin folds on the dorsum of the hand were used. A better system might have been to use radiographic or ultrasonic measurements of the breast skin at the test site. Changes in thickness during therapy could not be detected.

Effects of radiation on animal tissue have been widely reported. When the

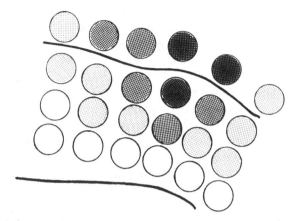

Fig. 32. Area extensions of skin about the knee joint when flexed through 90°; shading is in proportion to the degree of extension. The set of circles at the top represents the mid-line markings; those within the leg boundary, the lateral markings. Maximum extension (approaching 90% increase in area) is shown by the densest shading. Circles without shading represent minimal or negative area extension.

properties of rat skin were measured 50 and 120 days after a single dose of radiation, stiffness decreased with dose at doses of 1000 to 3000 rads, the drop being more noticeable at 120 days. Skin thickness measurements on these rats showed a slight thickening with increasing dose, but scanning electron microscopy showed matted collagen fibres suggesting that the changes may be due to alterations within the collagen or in the collagen–matrix interaction rather than in skin thickness (Ranu, Burlin & Hutton, 1975).

In general, radiation has little or no effect on thickness or mechanical properties of skin until quite high dose levels are attained, when interaction with the tissue components leads to breakdown of the structures.

SKIN STRETCH DURING NORMAL MOVEMENT

Extensions of the skin produced during normal physiological movement do not always correlate directly with the changes during controlled mechanical testing and have to be investigated directly. The knee joint provides a useful model; it is readily accessible and has a relatively simple motion. A two-dimensional 'map' of skin movement about the joint can be obtained by printing or lightly stamping a circle on the skin with the joint relaxed or in a pre-determined rest position. By repeating the printing, a grid can be laid out over the joint (Fig. 32) which enables the observer to see the relative differences in stretch immediately the joint is flexed.

Quantitative data can be obtained by measuring the diameter of the circles and the major and minor diameters of the distorted circles after flexion. By assuming that all the outlines are ellipses, the area may be determined and hence the change in area or area extension

Patterns of high and low extensions are normally found. The maximum extension values are generally on the central medial sets of outlines proximal to the upper border of the patella. Actual values of extensions vary widely from person to person and the maximum extension recorded in each subject varies from 35% to almost 100%. Although it is difficult to correlate this sort of data with physical typing, the maximum values occurred in tall, athletic subjects rather than obese types, indicating that muscle masses near joints and the size of the joint are probably the principal causes of large extensions (P. F. Millington & R. Wilkinson, unpublished data).

If a joint is fully flexed, regions of the skin along the principal flexion axis frequently show blanching. The imposed tissue pressure is thus sufficient to close down the upper dermal capillary network and maintenance of this position causes pain and possibly eventual tissue damage.

EFFECTS OF DRUGS

Many drugs affect connective tissues including skin. Lathyrogens, e.g. seeds of *Lathyrus odoratus* or β-amino-propionitrile, cause a pronounced loss of strength because of the failure of collagen to cross-link (Fry *et al.*, 1962; Friedrich, Wuppermann & Zimmermann, 1975). As a result the collagen remains in a highly soluble form and the dermis has little cohesion. Indeed all collagen-containing tissues are similarly affected.

Drugs which inhibit cellular proliferation, e.g. cyclophosphamide (Wie, Engesaeter & Beck, 1979) or D-penicillamine (Vogel, 1975; Friedrich *et al.*, 1975), may also reduce skin strength, often in proportion to the dose given. Antibiotics such as oxytetracycline (Engesaeter & Skar, 1978), cloxacillin and fusidic acid (Engesaeter & Skar, 1979) may likewise reduce skin strength.

That sex hormones affect skin is now generally recognised. Administration of male hormones to domestic fowl increases the tensile strength of skin and reduction of male hormones by caponising reduces skin strength. Oestrogen administered to male fowl also decreases skin strength (Herrick, 1945; Herrick & Brown, 1952). However, in studies on the indentation of human skin topical or parenteral testosterone had no effect on elderly men, but topical oestrogen reversed some of the age changes in elderly women (Chieffi, 1950). The effects of steroids are complex. Vogel (1972) administered prednisolone to rats, and after short-term administration found that when allowance for skin thickness has been made, strength was increased and the skin was stiffer. It has been suggested that this was due to increased collagen cross-linking. After longer administration, skin became thinner and its strength decreased. After treatment with cortisol, which has similar effects on mechanical properties, the ratio of insoluble collagen to the other fractions is increased and the skin is stiffer (Vogel, 1974). The strain-to-failure was slightly smaller in treated rats than in controls, but Burton & Shuster (1973) reported an increase in extensibility of human skin tested *in vivo* 2 hours after a single infusion of 2 g prednisolone over a period of 2 hours as a treatment for alopecia areata. Burton & Shuster (1973) also reported an increase in soluble and no change in insoluble collagen after glucocorticoid administration and suggested that prednisolone reduces cross-linking.

Not surprisingly, a simple list of steroid effects on skin in different species appears to give contradictory results and even to suggest different explanations. Indeed, Engesaeter & Skar (1979) have shown that changes in mechanical properties of tissues following drug administration vary from tissue to tissue in the same animal. Consequently, unless the full effects of a drug are known, great care must be taken to monitor changes in the skin and to remain open to the possibility of orthopaedic problems.

VISCO-ELASTIC PROPERTIES OF SKIN

So far the time dependence of skin has been considered only by specifying strain rate or rate of load increase. Another group of tests can be designed to investigate time dependence more explicitly; these measure creep and stress relaxation. In creep tests, a specimen is rapidly loaded to a pre-determined level and length measured as a function of time. In stress relaxation tests, on the other hand, the specimen is stretched to a given length and the load on the tissue measured as a function of time.

Difficulties arise from the fact that although modern testing machines are capable of loading or stretching tissue rapidly, the transition is not instantaneous. The consequences of finite transition times diminish as the test progresses. As a working rule, it is taken that at times greater than $10 \times$ the transition period the effects can be considered negligible. Unfortunately some of the interesting responses occur in the early part of the curve and it is necessary to know the capability of the system before interpreting the data. Most stress relaxation curves follow some sort of exponential decay and so the most rapid change occurs immediately after the initial stretching. So, if we assume a transition time of 1 second, then between 0.1 second and 0.2 second one-tenth of the strain has already been applied, and thus the initial stress relaxation of that strain will occur while this strain increment is being applied. The total load measured will thus be less if the transition took 1 second than if it had taken 0.1 second. Furthermore, for tissue, relationships between stress, strain and time cannot be assumed to be linear (or even simple), and therefore correction of data is at best approximate until long times have passed. Thus, if any modulus is to be computed, it is preferable to keep the transition time as short as the testing machine will allow.

Creep and stress relaxation are inversely related. If the material is 'linear' one can be computed from the other. Unfortunately this is not possible for skin since the relationships are not sufficiently well defined. Vogel (1973, 1977) studied both stress relaxation and creep in order to investigate the effects of various pharmaceutical products on the responses, but the strain rate used for the stress relaxation measurement was only 30 mm/min and that for creep a little better at 5 cm/min. Although his tests may show consistent differences between the effects of different pharmaceutical products, the results are not well suited for the calculation of moduli. On the other hand, Daly (1966) and Al-Haboubi (1977) carried out stress relaxation tests on human skin at strain rates of 50 cm/min and these results are among the best available.

Since creep and stress relaxation tests take much longer than tensile tests, control of the environment is even more important. Daly performed his tests in an immersion tank and Al-Haboubi was investigating the effects of immersion fluid on stress relaxation. Imbibition of fluid occurs even with balanced salt solutions and high-humidity chambers may be preferable if the internal loading of the specimen is not a problem.

Various ways of establishing standard parameters for visco-elastic behaviour have been suggested. None of them is entirely satisfactory, but one of the simplest is the 'normalised stress relaxation modulus' which is obtained by taking the ratio of the value of the stress after a pre-determined time (say 100 seconds) to the initial stress, to give '100-second modulus'. It is very important to use the same conditions for the test, especially the initial strain rate, before attempting to make comparisons between different specimens.

Another way to obtain 'normalised stress relaxation data' rests on the assumption that an exponential decay of stress allows the extrapolation of the plot to very long times. The difference in stress, $\sigma_0 - \sigma_\infty$, is the difference between the initial stress and the final stress at 'infinite time'; this may be normalised by dividing by initial stress, i.e. $(\sigma_0 - \sigma_\infty)/\sigma_0$. Unfortunately it seems that at very long relaxation times, skin may exhibit unexpected behaviour. Lanir (1976) observed an *increase* of stress in some specimens after 400–500 seconds. This phenomenon was also observed by Al-Haboubi and would seem to invalidate the use of measures involving σ_∞. No explanation for this behaviour has been suggested.

Similar problems exist for creep tests and analogous moduli may be computed. The major difference between the two tests is that creep tests may be terminated by rupture of the specimen and marked changes in cross-sectional area may occur.

TORSION TESTS

Because it is not possible to determine accurately the area of skin under test in studies *in vivo*, the simple stress relaxation and creep tests are not very meaningful under these conditions. However, torsion tests are suitable for the study *in vivo* of different strain rates and short-term stress relaxation or creep.

Early work was carried out by Vlasblom (1967) and his apparatus used later by Sanders (1973). A single disc was attached to the skin surface with double-sided adhesive tape and a constant torque applied for a set period of time. This can be regarded as a form of creep. The response was found to be of the form:

$$U(t) = U_{\mathrm{E}} + U_{\mathrm{V}} [1 - \exp (t/\tau)] + kt$$

where $U(t)$ is the response at time t, U_{E} the initial elastic response, U_{V} a time-dependent visco-elastic response, τ a constant having the dimensions of time, and k a constant.

Changes due to age were investigated and it was found that the total deflection at 2 minutes increased with age, mainly as a consequence of the initial response U_{E}. U_{V} and k showed a much smaller age dependence. Female subjects seemed to show a greater deflection than males of similar age, but the small number of tests did not permit statistical evaluation (Sanders, 1973).

A major defect of this early equipment was the ill-defined area under test. If a guard ring is placed outside the rotating area, then area definition is improved (Finlay, 1970). Cyanoacrylate adhesive was used for attachment since double-sided adhesive tape showed stress relaxation at the applied loads. The experimental procedure varied in that a set torsional strain was applied and the stress relaxation measured.

Using this arrangement, the low-load extensibility of skin decreased with age (Finlay, 1971), in apparent contradiction to Sanders' results. An explanation may be that in old age skin becomes flaccid and in Sanders' tests skin 'migrated' to increase the deflection. In Finlay's more constrained device, movement of skin outside the test area would not affect the data. Reduction in low-load extensibility would be in accordance with the evidence from uniaxial tests.

Torsion tests can also be used effectively for the investigation of hysteresis and strain rate or frequency effects. By applying sinusoidal displacements at various amplitudes, it can be shown that the relationship between torque and displacement is linear up to amplitudes of about $\pm 2°$. Above this amplitude, skin begins to stiffen, 'peaking' the curve (Fig. 33). In the frequency range 0.004–1 Hz, torque leads displacement by about $10°$; at 20 Hz the phase angle rises to $17°$. Frequencies above 20 Hz could not be reached by the apparatus (Finlay, 1978).

The relationship between torque and angular displacement can be expressed simply if the initial effects are neglected, thus:

$$\tau = k\theta + k_1 \cdot \mathrm{d}\theta/\mathrm{d}t$$

where τ is the torque and θ the angular displacement. The coefficients k and k_1 are not constant. The amplitude dependence of k can be readily demonstrated in variable-amplitude tests at frequencies below 1 Hz. The term $k_1 \cdot \mathrm{d}\theta/\mathrm{d}t$ is effectively constant and so over the range of frequencies up to 1 Hz, k_1 is inversely proportional to $\mathrm{d}\theta/\mathrm{d}t$. The coefficient would also be expected to vary with θ.

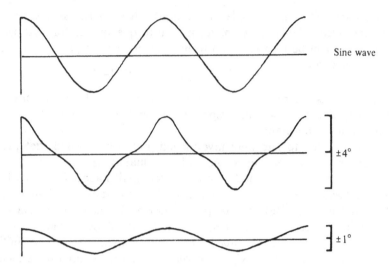

Fig. 33. Torque responses obtained for different values of sinusoidal displacement. Increasing skin stiffness is apparent in the $\pm 4°$ curve.

INDENTATION AND RECOVERY

When a small object is pressed into the skin and then removed, the resultant depression gradually recovers over a period of a few minutes. Experiments to measure these events and to establish parameters for quantifying indentation and recovery were first performed by Schade (1920). A modified version of his apparatus was used later to determine variations in these parameters with age (Kirk & Kvorning, 1949; Chieffi, 1950; Kirk & Chieffi 1962).

A 50-g load was applied to a small hemispherical indenter, measuring the movement relative to the skin surface. The weight was left in place for 2 minutes, removed and the recovery measured immediately and after 3 minutes. Skin over the medial aspect of the tibia was used as this has a firm, fairly flat support and contributions from soft connective tissue would be small. Four parameters were used to characterise the process: (*a*) the total depth of indentation after 2 minutes of applied load; (*b*) the initial depth of indentation; (*c*) the initial recovery immediately after removing the load; and (*d*) the amount of recovery after 3 minutes. In young people of 18 to 25 years it was found that the initial recovery of the skin and the total indentation were significantly greater than in people over 60 years (Kirk & Kvorning, 1949).

A further study over a greater range of ages showed that the percentage immediate indentation and the percentage immediate recovery taken as a proportion of total indentation decreased with age. The final percentage recovery figures decreased with age up to 40 years, but were relatively constant thereafter (Kirk & Chieffi, 1962).

A more sophisticated way to measure the elastic component of indentation recovery is to measure the coefficient of restitution. This coefficient is normally measured as the ratio of the rebound height to the initial height when a hard smooth object, often a ball, falls onto a resilient surface. Young skin has a higher coefficient of restitution than old skin, the difference being greater in areas exposed to the sun. Pathological conditions also affect the coefficient. It is noticeably lower in cases of atrophy, sclerosis, dermal infiltration and epidermal acanthosis, and significantly lowered 48 hours after death. The value of the coefficient is higher in cases of lymphangitis, lymphoedema, tension bullae and capillary angiomas (Tosti *et al.*, 1977).

EFFECT OF PATHOLOGICAL CONDITIONS

Gross changes in mechanical properties frequently accompany clinical conditions of skin, but the relationship between these changes and the disease is not always simple to determine. Factors such as changes in structure, changes in function and changes in anatomical composition may have a direct effect on the particular mechanical or physical property being measured, but the technique used must have the appropriate level of sensitivity for the expected change. Consideration of three disorders will illustrate these points.

EHLERS–DANLOS SYNDROME

The Ehlers–Danlos syndrome is a disorder characterised by highly extensible, but elastic skin. The disease exists in several forms with varying types of inheritance. In many patients it is possible to find localised areas of skin with extensions of 1000%.

A similar condition in calves, known as dermatosparaxis, involves a deficiency of pro-collagen peptidase, an enzyme necessary for converting pro-collagen to collagen. The effects of this deficiency are similar to those found in Ehlers–Danlos type VII (Pierard & Lapiere, 1976). In lambs which fail to survive because of extreme skin fragility the pro-collagen form is the major collagen component. Absence of retraction after extension seems to be due to an absence of linking between the pro-collagen fibrils, a condition not unlike that induced by lathyrogens in the foetus and young animals. Microscopic studies of such tissue revealed the presence of broken-ended connective tissue fibrils in both type I and type II cases (Black *et al.*, 1980). This is in sharp contrast to normal connective tissue where the ends of collagen fibres are not identifiable. If the tissues are not so disrupted then the relatively normal elastin fibres are capable of restoring the skin to its pre-deformation state, a situation which is typical of the usual human forms of the syndrome.

CUTIS LAXA

Cutis laxa is a rare disorder of connective tissue in which the skin hangs in loose folds. There are several variants of the condition, some inherited and some acquired. Most forms are relatively benign, but the recessive inherited form often has associated pulmonary and cardiovascular abnormalities, which may lead to death in childhood. Histopathological studies indicate a damaged elastic fibre network and it has been suggested that this causes the loose skin. Abnormal elastin would readily explain the pulmonary and vascular defects.

Studies using a suction cup technique failed to demonstrate any differences between six patients and age- and sex-matched controls (Grahame & Beighton, 1971). The methodology of this paper was criticised by Black & Shuster (1971), but in any case the loading curve, which is more dependent on collagen, would be less likely to give information than the unloading curve, which is more dependent on elastin. No reports on unloading were presented.

PSEUDO-XANTHOMA ELASTICUM

Patients with pseudo-xanthoma elasticum have small nodules on parts of their skin, primarily in the flexion creases. The uninvolved skin is macroscopically normal but histopathology reveals aggregations of elastin-staining material in the dermis and there are always changes in skin elasticity. Four types of the condition are recognised, two with dominant and two with recessive inheritance. The dominant groups have measurably lower and the recessive groups higher elastic moduli as measured by a suction cup technique (Harvey, Pope & Grahame, 1975).

Pathological changes in the fibrous extracellular components of skin may be causally related to the observed mechanical responses of the tissue. Failure to establish adequate cross-linking, as in Ehlers–Danlos syndrome type VII or lathyrism, is responsible for the reduced tissue strength. In severe cases the collagen may have so little cohesion that the elastin is unable to restore the original configuration, or the dermis becomes so fragile that even simple flexion of a joint could produce extensive fracture. On the other hand, if the elastic network is fragmented, as in cutis laxa, then the overall strength of skin may not be impaired, but retraction is grossly affected. The increased amount of elastic tissue present in pseudo-xanthoma elasticum is now known to affect mechanical properties directly even though the precise biochemical and stereological correlations have still to be established.

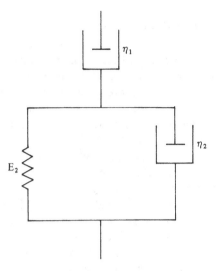

Fig. 34. Three-component viscous model for skin. The upper viscous element represents unrecoverable creep; the Voigt element (spring E_2 and dash-pot η_2) represents the recoverable creep.

MODELLING

Because the stress–strain response of skin is so difficult to characterise, it is impossible to compare specimens except in general terms. Various modelling techniques have been tried in an attempt to overcome these difficulties. These techniques may be divided into the physical, in which arrays of springs and dash-pots are used, and the mathematical.

PHYSICAL MODELS

Historically, physical models were seen as one of the simplest and most useful ways of modelling visco-elastic materials. Basically the models consist of springs, representing elastic response; dash-pots, representing viscous response; and sometimes dry frictional elements representing a 'switch-type response' in which no reaction occurs until a particular load has been reached. Arrays of components may be analysed using classical mechanics, and if the arrangements chosen mimic tissue behaviour over a reasonable range of conditions, differences recorded between tissue samples may be related to variations in the constants associated with the various components.

The simplest model used to simulate skin is the three-component arrangement used by Jamison, Marangoni & Glaser (1968) that is shown in Fig. 34. More complex models have been used, but they seem more successful for parallel-fibred tissue than networks such as skin.

MATHEMATICAL MODELLING

A theoretical description of skin, if available, would allow comparisons between specimens to be made in terms of the constants in the equation, comparisons between tests performed at different strain rates and behaviour in a particular test mode to be predicted from constants measured in a different mode. In the hope of developing such a model, numerous attempts have been made to develop mathematical expressions for some or all of the stress, strain and time relationships of skin.

Before discussing some of the approaches to mathematical modelling, it is necessary to outline some of the concepts and mathematical notation used. Until now, the conventional engineering description of stress has been used, i.e. force per unit area, F/A_0. This may also be termed Lagrangian stress, since Lagrange proposed the coordinate system whereby a particular point in the material is specified by its original position in the undeformed state. In a theoretical treatment, it is frequently more convenient to use the 'true', Cauchy or Eulerian stress, F/A, where A is the actual area at the time in question. In complex deformations, a third type of stress may be useful; this according to its precise formulation may be known as the Kirchhoff, Piola or Kirchhoff–Piola stress, where the force is measured as a vector in the coordinate system of the deformed body, but per unit area of the undeformed material.

The coordinate system X_A ($A = 1,2,3$) usually refers to material in the undeformed state and may be any useful system: Cartesian, cylindrical or spherical polars or general curvilinear. This last system is not in common use and may require explanation. If a system of three sets of curved surfaces is imagined in space such that at any point three surfaces (one from each set), intersect, then these may form a coordinate system. For a fuller account see Green & Zerna (1954). Since these surfaces are curved, the point at which they intersect may be described in two ways. The first uses the three vectors lying along the lines of intersection of pairs of surfaces, where g_1 lies along the intersection of the θ_2 and θ_3 surfaces and the set of vectors g_i ($i = 1, 2, 3$) are referred to as covariant. The second is written g^i where $g^1 = \epsilon_{rst}\, g_2 \times g_3$ is normal to the surface θ_1 at the point in question. ($\epsilon_{rst} = 1$ if rst are right-handed, -1 if left-handed and 0 if any two are equal.) The g^i set of vectors are referred to as contravariant (see Fig. 35).

The position vectors x_i usually refer to the spatial coordinate system and so a point X_A in the material may move from an initial position which may be written $_i x$ or x_0 (i) to a position x_i or $x(i)$. (If the coordinate system is the only set of subscripts or superscripts in use, then instead of using $i = 1, 2, 3$ the three axes may be specified as ijk).

Tensors may be covariant, contravariant or mixed. If the order of suffices in a mixed tensor is important, then 'stops' are used: e.g. if written $t^i_{.j}$ then $t^i_{.j}$ is not necessarily equal to $t_j{}^{.i}$ but if written t^i_j then $t^i_{.j}$ equals $t_j{}^{.i}$. Commas

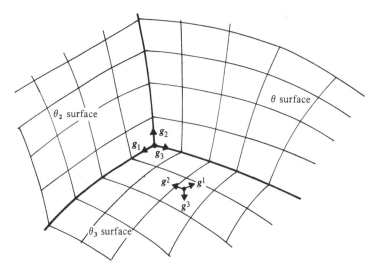

Fig. 35. Contravariant (g^i) and covariant (g_i) base vectors in a generalised curvilinear coordinate system.

may be used to denote differentiation: e.g. if $A_{jk} = f(\theta^i)$ then $A_{jk,i} = \partial A_{jk}/\partial \theta^i$. Repeated suffices imply summation: e.g. $x_i x_i = x_1{}^2 + x_2{}^2 + x_3{}^2$. This notation is very compact, but requires careful thought and it is easy to misread an equation. However, many authors generalise their formulae in this way and it is hoped that this will assist in translation.

Strain is more difficult to define mathematically. The common engineering strain for uniaxial extension is $(l - l_0)/l_0$, but a variety of other measures involving the initial length and final length have been proposed, some only suitable for very small or infinitesimal strains. To help those who wish to begin using these expressions, a list of useful formulae will be found at the end of the chapter for both uniaxial extension and, where appropriate, the full strain tensor for complex deformations. It should be noted that for a given initial and final length the numerical value of strain calculated by the different measures is widely different and so the measure used should be recorded and reported carefully.

For the purpose of our discussion, the models may be divided into three categories: network analysis, continuum models assuming the existence of a strain energy function, and those continuum models in which such an assumption is not made. The assumptions made in each model are stated and it should be noted that in most cases at least one assumption is made which is dubious in order to make the mathematics more tractable, and so the lack of agreement with the experimental results should not be too surprising.

It is important to note that in order to enable readers unfamiliar with the mathematical expressions to follow this section more readily, all symbols have been standardised and so the equations may not be identical to those in the

original papers. To help further, a list of notation and definitions will be found at the end of the chapter.

NETWORK ANALYSIS

One of the early forms of network analysed was the rhomboidal mesh proposed by Gibson *et al.* (1965). The rods forming the mesh are rigid, but with compressible links; light springs attach at the corners to supply the necessary restoring force. A plot of a computer simulation was given with experimental points, but no details of calculation presented. Unfortunately the drawing of the model as presented is a geometric impossibility.

A very simple network has been used to predict strain at entry to Phase II (Markenscoff & Yannas, 1979). If a single collagen fibre is assumed to consist of a number of short sections of unit length, lying in a plane, which transmit load efficiently at the section ends and where the angle between the section and the vertical varies randomly from 0° to 180°, then there will be a monotonic progression of the fibre from left to right. The expected length of the fibre may be written

$$E(l) = n/\pi \int_0^\pi \sin \theta = 2n/\pi$$

where there are n sections per fibre.

If the strain to straighten the fibre is ϵ_s then

$$E(l)(1+\epsilon_s) = n$$

also

$$(2n/\pi)(1+\epsilon_s) = n$$

therefore

$$\epsilon_s = (\pi-2)/2$$

$$\epsilon_s = 0.57 \text{ or } 57\%$$

This value is similar to that at entry to Phase II found for animal and young human skin. In older people θ is believed to lose its random nature. The model could be adjusted for anisotropy by introducing a different probability function for θ.

A model based on 'collagen arrival density' was proposed by Soong & Huang (1973). Only fibres oriented in the direction of load are considered to participate in load resistance. The probability of any one fibre participating can be written as $P(x)dx$, the collagen arrival density.

When l_s is defined such that

$$l_s + 0.1 \, l_s = 2l \quad \text{(see Fig. 36)}$$

this assumes an extension of 100% for elastin before fracture and 10% for collagen.

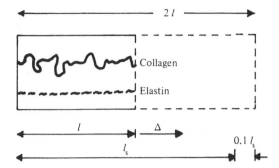

Fig. 36. Stochastic model of skin with three components: (*a*) homogenous soft matrix–ground substance, (*b*) collagen fibres, of length l_s when straightened, (*c*) elastin fibres. The specimen (length *l*) is elongated by an amount Δ such that each collagen fibre is straightened and then extended 0.1 of its total original length, l_s. Thus: $l_s + 0.1\,l_s = 2l$.

If
$$y = (l+\Delta) - l_s$$

and
$$z = y/l = (1+\Delta/l) - 2/1.1$$

so that *y* is the difference between the extended specimen $(l+\Delta)$ and l_s, the bounds of interest in this model are from $-9/11$ to $2/11$. The model arrives at two predictions of the tangent modulus (Phase III modulus) $Y_t(x)$:

$$E[Y_t(x)] = Y_e\left[V_{ce} - B\int_{-9/11}^{x} P(u+2/11-x)\,du\right]$$
$$+ Y_c B\int_{-9/11}^{x} e^{au}P(u+2/11-x)\,du \quad (-9/11 \leqslant x \leqslant 0)$$

and

$$E[Y_t(x)] = Y_e\left[V_{ce} - B\int_{-9/11}^{x} P(u+2/11-x)\,du\right]$$
$$+ Y_c B\int_{-9/11}^{0} e^{au}P(u+2/11-x)\,du$$
$$+ Y_c B\int_{0}^{x} P(u+2/11-x)\,du \quad (0 \leqslant x \leqslant 2/11)$$

$$B = \Psi_c V_c \Big/ \int_{-9/11}^{2/11} P(u)\,du$$

where *Y* is the modulus, *V* the volume, *x* a spatial coordinate, *a* is a constant, subscript t refers to the tangent modulus, e to elastin and c to collagen, Ψ_c is the percentage of collagen fibres and *u* is a generalised form of *x*.

Perhaps the most ambitious model dealing with flat homogeneous tissues is that of Lanir (1979). His basic assumption is of undulating collagen fibres cross-linked at intervals by thin, straight elastin fibres. Two situations are discussed in the paper, one having a high and the other a low density of cross-links. He suggests that in the initial phase of extension the segments of collagen between cross-links may be divided into two groups: those in which the collagen has not yet straightened and in which only elastin carries load and those where the collagen has straightened and both fibres carry load. Thus, in principle, the stress–strain curve could be computed from a knowledge of the spring constants of the two fibres and the probability of any given segment becoming straightened.

The following model is an attempt to predict the results of a quasi-static biaxial tensile test with a high density of cross-links. Using the elastin fibre to define the direction of the segment a probability function $P(\theta)$ can be defined such that

$$\int_{\pi/2}^{\pi/2} P(\theta)\, d\theta = 1$$

If a line of unit length is drawn at an arbitrary angle μ then the number of fibres lying between θ and $\theta + d\theta$ will be $NP(\theta)\, d\theta$, where N is a measure of fibre density and the number of segments crossing this line $NP(\theta)\, d\theta \cdot \cos(\mu - \theta)$. Each of these segments will stretch and rotate and this change can be written in terms of the extension ratios in the principal axes and a shear ν.

The force, T_{ns}, exerted by the segments crossing a unit vector n in the direction s, may be written as

$$T_{ns} = \int_{-\pi/2}^{\pi/2} W_1(\lambda, \theta) \cos(\mu - \theta)[\cos\psi\, (\lambda_1 \cos\theta + \nu\lambda_2 \sin\theta)$$

$$+ \sin\psi \cdot \lambda_2 \cdot \sin\theta]\, d\theta + p(\lambda_1, \lambda_2, \nu) \cos(n_1 \cdot s)$$

where $W_1(\lambda, \theta) = F(\lambda)/\lambda$, $F(\lambda)$ is the tension in the fibre at extension ratio λ, μ is the angle of n to the axis, ψ the angle of s to the axis and $p(\lambda_1, \lambda_2, \nu)$ the hydrostatic pressure at the given state of extension.

The hydrostatic pressure in the matrix, p, is small at physiological stresses, but may become significant at high stress. The model successfully predicts stiffening as extension progresses and a suitable form of $P(\theta)$ will allow anisotropy. Techniques for the determination of $P(\theta)$ in real tissues are discussed in the original. While this model has enough flexibility to account for a number of situations in quasi-static mechanical tests, it takes no account of time in any of its predictions.

STRAIN ENERGY FUNCTIONS

The strain energy, W, can be taken as a function of the state of deformation only. This applies only to elastic materials. If irreversible deformations occur then W is properly termed a work function and other variables may be involved. Strain energy is normally expressed in terms of three functions of the extension ratio, λ, split into its Cartesian coordinates.

These functions are the strain invariants I_1, I_2, I_3:

$$I_1 = \lambda_1{}^2 + \lambda_2{}^2 + \lambda_3{}^2$$
$$I_2 = \lambda_1{}^2\lambda_2{}^2 + \lambda_2{}^2\lambda_3{}^2 + \lambda_3{}^2\lambda_1{}^2$$
$$I_3 = \lambda_1{}^2\lambda_2{}^2\lambda_3{}^2$$

If $I_3 = 1$ the material is incompressible and even if this assumption is not stated explicitly, derivations in which $I_3 = 1$ only apply to incompressible materials. In the general case

$$\sigma_i = \frac{2}{\lambda_i}\left[\lambda_i{}^2\frac{\partial W}{\partial I_1} + I_2\frac{\partial W}{\partial I_2} - \frac{I_3}{\lambda_i{}^2}\frac{\partial W}{\partial I_2} + I_3\frac{\partial W}{\partial I_3}\right]$$

and so if W is known then the material can be characterised. When fixing numerical values, it may be useful to remember that $W(3,3,1) = 0$, since it is assumed that there is zero strain energy at zero deformation.

The case of an isotropic solid under uniaxial tension was considered by Veronda & Westman (1970), where $\lambda_2 = \lambda_3$.
Then

$$\frac{\sigma}{2(\lambda_1 - \lambda_2{}^2/\lambda_1)} = \frac{\partial W}{\partial I_1} + \lambda_2{}^2\frac{\partial W}{\partial I_2}$$

If the material is incompressible so that $I_3 = 1$ and

$$\lambda_2{}^2 = \lambda_3{}^2 = 1/\lambda_1$$

then

$$\frac{\sigma}{2(\lambda_1 - 1/\lambda_1{}^2)} = \frac{\partial W}{\partial I_1} + \frac{1}{\lambda_1}\frac{\partial W}{\partial I_2}$$

Veronda & Westman selected, on the basis of previous work, a polynomial form of W and finally suggested

$$W = c_1[\exp(\alpha(I_1 - 3)) - 1] + c_2[I_2 - 3] + g(I_3)$$

where c_1, c_2, α are constants and $g(1) = 0$.
Curve fitting was performed and yielded

$$W = 0.00394\,[\exp(5.03(I_1 - 3)) - 1] - 0.01985\,(I_2 - 3) + g(I_3)$$

The authors felt that a better fit would be obtained if $g(I_3) \neq 0$, i.e. the material was compressible, but the accuracy of their experimental data did not allow $g(I_3)$ to be determined and the possibility that $g(I_3) = 0$ could not be excluded. This would seem to imply that c_1 and c_2 may have been quoted to more places than the accuracy of the data would justify. It should be noted that North & Gibson (1978) claimed to show that human skin is in fact incompressible.

If the strain conditions are defined as $\lambda_1 = \lambda$, $\lambda_2 = 1/\lambda^{\frac{1}{2}} = \lambda_3$, termed homogeneous deformation by Hildebrandt, Fukaya & Martin (1969), then for an incompressible material

$$I_1 = \lambda^2 + 2/\lambda$$
$$I_2 = 2\lambda + 1/\lambda^2$$

and the expression for σ reduces to

$$\frac{\sigma}{2(\lambda^2 - 1/\lambda)} = \frac{\partial W}{\partial I_1} + \frac{1}{\lambda}\frac{\partial W}{\partial I_2}$$

which, if W is only dependent on λ, could be written as

$$\sigma = (\lambda^2 - 1/\lambda)f(\lambda)$$

For soft tissue such as mesentery they propose

$$f(\lambda) = \frac{k_1}{(\lambda_{max} - \lambda)^a} + \frac{k_2}{(\lambda - \lambda_{min})^b}$$

where λ_{max} and λ_{min} refer to the Phase II transitions in extension and compression respectively.

One of the difficulties in dealing with biological tissues is that the internal stress is not zero and when tissues are cut, retraction occurs. The true zero point is difficult to determine and there is frequently a small, unknown stress, σ', on the specimen at the start of mechanical testing. Snyder (1972) attempted to show how, using successive approximations, a numerical solution approximating to the true zero could be produced.

From the equation introduced by Fung (1968):

$$d\sigma/d\lambda = a\sigma + k$$

Integration yields $\sigma = k/a\,[\exp(a(\lambda - 1)) - 1]$

If variables marked with a prime, e.g. l_0' refer to measurements at the start of the test, variables marked with a double prime refer to measurements after some increase in stress σ'' over the unknown stress σ', and unmarked variables to the conditions at the true zero then

$$l_0' = \lambda' l_0$$

so

$$\lambda' = 1 + 1/a\,\ln[(a\sigma'/k) + 1]$$
$$\lambda = l''/l_0 = \lambda'\,l''/l_0' = \lambda'\,\lambda''$$

If $\sigma'' \gg \sigma'$ then $\sigma'' \simeq k/a\,[\exp(a(\lambda - 1)) - 1]$
Substituting for λ then λ',

$$\sigma'' \simeq k/a\,[\exp(a(\lambda'' - 1))\,\exp\{\lambda''\ln((a\sigma'/k) + 1)\} - 1]$$

thus $\sigma'' \simeq k/a\,[\exp(a(\lambda'' - 1))\,\{(a\sigma'/k) + 1\}^{\lambda''} - 1]$

Snyder published the equation as

$$\sigma'' \simeq k/a\,[(a\,\sigma' + 1)^{\lambda''}\exp(a(\lambda'' - 1))]$$

which either includes an unannounced approximation or an error in transcription during printing.

A strain energy, $\quad W = \dfrac{k}{a^2} \cdot \exp\,(a(\beta-1)) - k\dfrac{\beta}{a}$

where β is analogous to an extension ratio for complex deformations, was proposed by Snyder. Where the tissue is in uniaxial tension λ may be substituted for β. For a homogeneous, isotropic, incompressible solid then β must be a single valued function of the first two strain invariants. Thus, for uniaxial tension the stress–strain relationship is

$$\sigma = \left[\lambda_1 - \frac{1}{\lambda_1^2}\right]\left[\frac{\partial W}{\partial I_1} + \frac{1}{\lambda_1}\frac{\partial W}{\partial I_2}\right]$$

A different form of strain energy was proposed by Demiray (1972):

$$W = k/2a\,[\exp\,(a(I_1-3)) - 1]$$

which is similar to the first term of the form proposed by Veronda & Westman (1970). Using the Finger strain tensor and Cauchy stress the equation for the uniaxial extension of a cylindrical rod may be written

$$\sigma = \pi r_0^2 k\,(\lambda - 1/\lambda^2)\,\exp\,[a(\lambda^2 + (2/\lambda) - 3)]$$

for an incompressible, homogeneous, isotropic material. In the more general case

$$\sigma^{kl} = pg^{kl} + \phi C^{kl} + \Psi B^{kl}$$

where p is a boundary function and g^{kl} is the reciprocal metric tensor of x_k, $\phi = 2\partial W/\partial I_1$, $\Psi = 2\partial W/\partial I_2$, $B^{kl} = I_1 C^{kl} - C^k{}_m C^{ml}$. Tong & Fung (1976) point out that the energy functions are not strictly speaking thermodynamic energy functions since tissues are not perfectly elastic and the mechanical response depends on the strain history. However, they claim that the assumption may still give useful results. Using Kirchhoff's stress and Green's strain tensor

$$W = (\tfrac{1}{2}\alpha_1 e_{11}^2 + \alpha_2 e_{22}^2 + \alpha_4 e_{11} e_{22}) + \tfrac{1}{2}c\,\exp\,(a_1 e_{11}^2 + a_2 e_{22}^2$$
$$+ a_3 e_{12}^2 + 2a_4 e_{11} e_{12} + \nu_1 e_{11}^3 + \rho_2 e_{22}^3 + \rho_4 e_{11}^2 e_{22} + \rho_5 e_{11} e_{22}^2)$$

where α, a, ρ and c are all constants. Thus in principle there are 12 constants to be determined – a daunting task. This is still a simplified equation since it is for an orthotropic two-dimensional material. Some of the constants could be evaluated by carefully selecting the mechanical conditions. In a tensile test, the shear strains are absent, thus terms including N are zero. But in a constrained tensile test (one in which lateral contraction is inhibited) some shear exists. In two experiments on pre-conditioned rabbit skin with the directions of extension mutually at right angles, curve fitting with the approximation
$$\rho_1 - \rho_2 = 0, \quad \rho_4 = \rho_5$$

can be performed using a complex iterative procedure. The values of the constants are very dependent on strain history and Tong & Fung (1976) were unable to use their data to investigate changes due to other independent variables.

This illustrates a basic dilemma of mathematical modelling: if the model is sufficiently complex to provide a really good fit for any given mechanical test, then the values of the constants seem to vary too widely between control specimens to permit systematic investigations of other independent variables, while if the fit is not so good, then the applicability of the model and especially its underlying assumptions are brought into question.

Three classical approaches to continuum mechanics are discussed by Dehoff (1978). In all three, equations are derived for the stress relaxation of tissues and the constants used to predict the response in a constant strain rate uniaxial tensile test. It is shown that, although by assuming certain forms for different generalised functions all three approaches can be made to give the same equation for stress relaxation, different equations for tensile tests are produced.

Haut & Little (1972) started with the 'Fung equation' in the form

$$\sigma(t) = \int_0^t G(t-\tau) \frac{d\sigma^e}{d\lambda} [\lambda(\tau)] \frac{d\lambda(\tau)}{d\tau} d\tau$$

$G(t)$ is a normalised stress relaxation function where $G(0) = 1$, and τ is a time constant. This equation is only strictly applicable to an instantaneous extension, but may be used with high strain rates if the material is not strain-rate-sensitive at these speeds. For small strains, ϵ may be used instead of λ and σ^e written as $c\epsilon^2$, if $G(t) = a \ln t + b$. Then

$$\sigma(t) = Y\epsilon_0^2 [\mu \ln t + 1] \tag{1}$$

Where the equivalent modulus, Y, equals bc and μ equals a/b. For uniaxial tensile tests $\epsilon = \beta t$ and

$$\sigma(t) = Y\epsilon^2/2 [1 + \mu (\ln (\epsilon/\beta) - 3/2)] \tag{2}$$

In characterising the visco-elastic properties of polymers, Lianis (1963) derived the following equation for uniaxial stress relaxation:

$$\sigma(t)/(\lambda^2 - 1/\lambda) = [a + b + 2\phi_1(t) - 2\phi_3(t)] + 1/\lambda[2b + c + 2\phi_0(t)$$
$$+ 2\phi_2(t) + 2\phi_3(t)] + [b + 2\phi_2(t) + \phi_3(t)] [\lambda^2 - 1]$$

where ϕ_i is a relaxation function.

This equation may be simplified for biological tissues, where there is no limiting stress, by writing $a = c = 0$. If ϕ_2 and ϕ_3 are assumed to be identically zero, then for small strains

$$\sigma(t) = 6\epsilon_0 [\phi_0(t) + \phi_1(t) - \phi_0(t)\epsilon_0]$$

This equation reduced to equation (1) above if

$$\phi_1(t) = -\phi_0(t) = Y/6 [\mu \ln t + 1]$$

but for the constant strain rate test

$$\sigma(t) = Y\epsilon^2 \{1 + \mu [\ln(\epsilon/\beta) - 1]\}$$

Bernstein, Kearsley & Zapas (1963) predict that

$$\sigma(t) = \int_{-\infty}^{t} \left[\frac{\lambda^2(t)}{\lambda^2(\tau)} - \frac{\lambda(\tau)}{\lambda(t)} \right] h \left[\frac{\lambda(t)}{\lambda(\tau)}, (t - \tau) \right] d\tau$$

For uniaxial stress relaxation then

$$\sigma(t)/(\lambda^2 - 1/\lambda) = H(\lambda, t)$$

and if
$$H(\lambda, t) = 2\phi_1(t) + 2\phi_0(t)/\lambda$$

then once again equation (1) is produced. However, the constant strain rate substitution now yields

$$\sigma(t) = Y\epsilon^2 \{1 + \mu [\ln(\epsilon/\beta) - \tfrac{1}{2}]\}$$

Thus, even though the different approaches may yield the same equations for stress relaxation, the predictions for uniaxial tensile tests will be different, although the numerical values of the last two equations are similar for small strains.

From this, it may be seen that a generalised constitutive equation for skin allowing for structural variations, anisotropy, and all the other experimentally determined features would probably be mathematically intractable.

NOTATION

A	area	S_{KL}	Kirchhoff–Piola stress type II
A_0	initial area	s_{Ki}	Kirchhoff–Piola stress type I
a	constant	t	time
b	constant	V	volume
c^{kl}	Finger strain tensor	V_c	volume fraction of collagen
c	constant	V_e	volume fraction of elastin
c_{ik}^{-1}	inverse Cauchy deformation	V_{ce}	volume of fraction of elastin plus collagen
$d()$	differential		
$\partial()$	partial differential	W	work function, strain energy
$E()$	expectation	X_K	Lagrangian or material coordinates
e	strain – as specified		
F	force	x_k	spatial coordinates
$G()$	normalised relaxation function	Y	modulus
G^{KL}	reciprocal metric tensor of X_K	Y_t	tangent modulus
g^{kl}	reciprocal metric tensor of x_k	Y_c	modulus of collagen fibres
I	strain invariant	Y_e	modulus of elastin fibres
k	constant	α	constant
l	length	β	complex deformation analogue of λ the extension ratio
l_0	initial length		
n, N	number	ϵ	engineering strain
$P()$	probability	θ	angle
P, p	pressure	θ_i	curvilinear coordinate surface
p	boundary function	λ	extension ratio

ν	shear	τ	torque
ρ	constant	ψ	angle
σ	Lagrangian or engineering stress	Ψ_i	percentage
σ	True, Cauchy or Eulerian stress	Ψ_c	percentage collagen fibres
τ	time		

SUMMARY OF FORMULAE

Engineering or Langrangian stress $\sigma = F/A_0$

'True', Cauchy or Eulerian stress $\sigma = F/A$

Piola or Kirchhoff–Piola stress type I $s_{Ki} = J\,X_{K,j}\,\sigma_{j^2}$

Piola or Kirchoff–Piola stress type II $S_{KL} = X_{L,i}\,s_{K,i}$

Extension ratio $\lambda = l_1/l_0$

Engineering or Cauchy strain $\epsilon = (l_1 - l_0)/l_0$ (finite strain)

Green, St Venant or Lagrangian strain $= \frac{1}{2}\,[(l_\mathrm{r}/l_0)^2 - 1]$

$$\text{or } 2_{rs}e = g_{\alpha\beta}\,(r, x^\alpha)\,(s, x^\beta) - {}_{rs}g$$

True or Hencky strain $= \ln\,(l_1/l_0)$ (finite strain)

Almansi–Hamel strain $= \frac{1}{2}\,[1 - (l_0/l_1)^2]$

$$\text{or } 2e_{rs} = g_{rs} - {}_{\alpha\beta}g\,({}^\alpha x, r)\,({}^\beta x, s)$$

Swainger strain $= 1 - (l_0/l_1)$

Finger strain tensor $= C^{kl} = G^{KL}\,(x^k{}_{,K})\,(x^l{}_{,L})$

Poisson's ratio $= -$lateral strain/axial strain for uniaxial tension

Strain invariants $I_1 = \lambda_1{}^2 + \lambda_2{}^2 + \lambda_3{}^2$

$\qquad\qquad\qquad I_2 = \lambda_1{}^2\lambda_2{}^2 + \lambda_2{}^2\lambda_3{}^2 + \lambda_3{}^2\lambda_1{}^2$

$\qquad\qquad\qquad I_3 = \lambda_1{}^2\lambda_2{}^2\lambda_3{}^2$

Inverse Cauchy deformation tensor $C_{ik}^{-1} = (x^i{}_{,k})\,(x^k{}_{,K})$

Young's modulus $Y = \sigma/\epsilon$ in the elastic region

5

MECHANICAL, THERMAL AND ELECTRICAL PROPERTIES

MECHANICAL PROPERTIES OF SKIN COMPONENTS

There are five main components of skin: collagen, elastin, keratin, ground substance and cells. In the previous chapter the mechanical properties of the whole tissue were considered, but now the properties of isolated components are explored. Because it may be difficult to isolate particular tissue components, it is common practice to start with tissues where one component is more abundant.

With the exception of the epidermis, cells in skin either form focal structures localised in the dermis or exist as single or small groups of cells (e.g. fibroblasts). Tests on viable and non-viable skin indicate little or no contribution from the cells to the mechanical properties, unless lysis has occurred when the enzymes may affect integrity of the structures. Thus, with the exception of the epidermis, few reports can be found on dermal cells. The special contributions of structures such as hair follicles have been assessed by Brown (1971).

COLLAGEN

Collagen is the most abundant fibrous component of skin and its role in all connective tissues appears to be that of the major stress-resisting element. The isolation of collagen from other components of the tissue inevitably changes some of its properties. Tendon, especially rat tail, provides a ready source of fibres that are readily separated. Collagen in tendon exists in various stages of organisation: bundles of fibres which are themselves composed of fibrils in turn formed from groups of triple-helix collagen molecules. Ground substance, mainly composed of mucopolysaccharides, intersperses these structures and may be removed, in part, by incubation with an appropriate enzyme (hyaluronidase or α-amylase) or by a chelating agent (EDTA). Incubation of tendon in buffered isotonic saline will also remove much of the ground substance and eventually disintegrate the collagen structure. Mechanical properties are influenced by the interaction between ground substance and collagen (Minns, Soden & Jackson, 1973). However, the effects of these pre-treatments on the collagen fibre itself are less clear. Sugar-splitting enzymes should not alter the strength of the basic collagen fibril but Rao (1972) reported that collagen fibres teased from bovine skin displayed only

50% of their untreated strength when treated with α-amylase. Minns *et al.* were unable to show a statistically valid decrease in strength of tendon collagen after this treatment, although differences of up to 38% have been reported. The purity of the enzyme preparation is important since even small amounts of contamination with proteases will significantly affect results.

If skin is used as a source of collagen fibres, not only ground substance but also elastin must be removed. This has been attempted by incubation of a small specimen with elastase following the incubation with α-amylase or hyaluronidase to remove mucopolysaccharides. Hydroxyproline in the incubation medium indicates removal of collagen, but soluble collagen will always enter even non-active solutions and so the relative levels of hydroxyproline should be compared to ascertain the degree of collagen breakdown.

Alternatively collagen fibres may be micro-dissected from surrounding tissue. Avoidance of strain is important. Stain affinity to the Masson Trichrome is altered after deformation into Phase II, the collagen retaining a red colouration (Craik & McNeil, 1965). This stain reaction change is permanent after death until the tissue eventually breaks down. However, some delayed elastic recovery does take place even in experiments *in vitro* and so specimens should be left for a short time before mechanical testing.

Collagen fibres are seldom straight, but their precise configuration has still to be established. Diamant *et al.* (1972) suggested a 'crimped' configuration for collagen fibres in tendon, but this has been disputed by Evans & Barbenel (1975) who proposed a helical form. In skin, the 'helical' appearance of collagen is readily perceived in young tissue but in aged specimens the individual bundles are not so well marked and when identified appear to be straighter.

It is generally assumed that when stress is applied to a single fibre, it gradually straightens. This phenomenon occupies Phase I of the stress–strain curve as shown in Fig. 28. Using fibres derived from rat tail tendon, Diamant *et al.* (1972) found the entry into Phase II occurring at strains from 1.5% in the old to 13% in the young, with the mature adult at about 4–5%. A mathematical model of the stress–strain response has been based on 'elastica' theory. The material at the apex of the 'crimp' is assumed to be rigid compared with the material along the length of the segments and this material deforms in a manner similar to that of a cantilever beam (Fig. 37).

The response obeys the equation

$$e = (\sigma/Y) = e_\infty - \psi(Y/\sigma)^{\frac{1}{2}}$$

where e is extension, σ the applied tensile stress, and Y the modulus. The values of e_∞ and ψ are governed by the geometry of the system such that

$$e_\infty = \sec\theta_0 - 1$$

and

$$\psi = \frac{d(1-\cos\theta_0)}{l \quad \cos\theta_0}$$

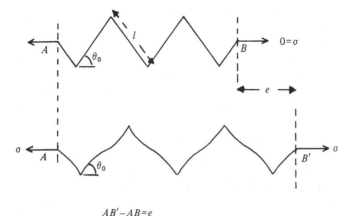

Fig. 37. Extension of a zig-zag model in which the apex points have infinite rigidity.

where l is the length and d the thickness of a segment, θ_0 is the crimp angle (see Fig. 37). The crimp angle seems to vary with age from 20° in the young to 12° in the old. l and d also vary with age: l varies from 21 ± 1 nm in the young to 110 ± 5 nm in the old and d from 100 ± 50 nm in the young to 510 nm in the old. Thus, the load-resisting unit in the newborn seems to be a single fibril, but several fibrils seem to act together in the old; this probably represents the increased cross-linking of collagen with age.

In the 'elastica' model the ratio of length to diameter seems to be about 5 and Comninou & Yannas (1976) feel that this is too low for the elastica theory, which implies a long flexible segment. Rather than assuming variable rigidity, as in the elastica model, they suggest a sinusoidal fibre geometry. The model could, however, be adapted to any regular waveform. In the final expression it is necessary to distinguish between the apparent strain (the movement of the fibre ends compared with the distance between the ends) and the material strain (the increase in length measured along the fibre compared with the initial length measured in the same way). For an initially straight fibre the two measures are necessarily identical. If the fibre geometry follows the general equation

$$y = a \sin bx$$

where the mean direction of the fibre lies along the x-axis then

$$\bar{\epsilon} = \epsilon + \frac{a^2 b^2}{4} \frac{\Lambda (\Lambda + 2)}{(\Lambda + 1)^2}$$

where

$$\Lambda = \frac{4\epsilon}{b^2 Q^2}, \quad Q^2 = \frac{4I}{A} = \frac{4D}{YA}$$

and $\bar{\epsilon}$ is the apparent strain, ϵ the material strain, I the moment of inertia of the cross-section, D the bending rigidity, and Y the elastic modulus in tension. This equation will simulate a three-phase curve of the type shown in Fig. 28 using suitable values for the constants.

Alternative mathematical models utilise the continuum mechanics approach already discussed in the previous chapter.

Since collagen fibres are composed of fibrils surrounded by an aqueous gel of mucopolysaccharide the interaction between fibril and mucopolysaccharide will affect the mechanical properties. Extraction of mucopolysaccharide affects the mechanical properties of tendon (Minns *et al.*, 1973) and fibres from bovine skin (Rao, 1972). Cohen, Hooley & McCrum (1976) suggest that the visco-elasticity of collagen is due to stressing of this interfibrillar gel. Such a mechanism would be influenced by the activation energy, ΔH, for viscous shear in the gel, while if the interaction is low, visco-elasticity would be governed by entropy changes, as exemplified in many rubbers.

An early attempt to investigate this interaction used samples of rat tail tendon tested at various temperatures between 2 and 32 °C, but no change in the mechanical properties with temperature could be detected (Rigby *et al.*, 1959). The properties of collagen thus seemed to be governed by entropy changes. This procedure would be unable to detect small mechanical changes because different samples were used for each test and inter-sample variability would be too great. The procedure was refined by Cohen and his colleagues who changed the temperature very rapidly during a creep test and measured the creep rates at the two temperatures just before and just after the change. If the instantaneous creep rates at the two temperatures T and T_0 are $\dot{e}(T)$ and $\dot{e}(T_0)$ then

$$\frac{\dot{e}(T)}{\dot{e}(T_0)} = \exp\left[\frac{\Delta H}{R}\left(\frac{1}{T_0}-\frac{1}{T}\right)\right]$$

or

$$\ln\left(\frac{\dot{e}(T)}{\dot{e}(T_0)}\right) = \left(\frac{\Delta T}{RTT_0}\right)\Delta H$$

where $\Delta T = T - T_0$ and R is the gas constant. From a series of experiments it was found that ΔH equals 12 kcal/mol for mature human tendon. Values of 20 kcal/mol and 33 kcal/mol were calculated for bovine flexor tendon and a reconstituted bovine collagen specimen respectively. These results indicate that interactions between mucopolysaccharides and collagen fibrils influence mechanical responses and that variations due to species, tissue and preparation method are probable.

There is obviously scope for more studies relating stress and strain in fibres, remembering that fibres are aggregates of fibrils, and also in relating the observations on single fibres to tissues. It would seem useful to establish whether collagen fibres from different tissues in the same species or the same tissue in different species do differ in their mechanical response or thermodynamic constant and to investigate the detailed effects of hydration on the mucopolysaccharide–collagen interaction.

ELASTIN

The definition of the term 'elastin' significantly affects the interpretation of the mechanical properties of tissue. Biochemically, elastin can be characterised by its amino acids, but by this stage the native conformation has been lost. Histologically elastin is characterised by its ability to react with certain chemicals, e.g. orcein. When examined at high resolution several components seem to be present. The core of the fibre is granular when examined in the transmission electron microscope and has been labelled 'elastin'. There are also the so-called micro-fibrils whose proportion diminishes with age, and finally fibrils of collagen which appear to wind around the elastin core in such a way that at extension ratios of about 1.3 the collagen fibrils themselves are straightened and stressed. Variations in the properties of specimens under test may be due to variations in the proportion of these constituents or to changes in the elastin core due to interactive preparation procedures or hydration.

The most common method of preparation is repeated autoclaving which denatures the other materials leaving the insoluble elastin plus some secondary components. This technique is most successful on ligamentum nuchae which has a very high proportion of histologically identified elastin. It will not completely remove the non-elastin components from lung. A more vigorous extraction uses 0.1 M sodium hydroxide at a temperature of 98 °C and this will produce a material having the chemical characteristics of elastin but which has an altered secondary structure (Partridge, 1962).

Many other techniques have been tried based on protein denaturing agents with and without enzymatic digestion. Ross & Bornstein (1969) started with 5 M guanidine chloride in a solution of 0.05 M dithioerythritol buffered with 0.1 M TRIS to pH 8.5 and 0.1% EDTA. The suspension of tissue was centrifuged and the pellet then incubated with collagenase. Szigeti *et al.* (1972) used 2.7% trichloroacetic acid at 90 °C for 30 minutes followed by four extractions in 8 M urea with 0.1 M 2-mercaptoethanol. Formamide has been used but has been shown to affect the mechanical properties of elastin prepared from ligamentum nuchae by repeated autoclaving (Mukherjee, 1969).

The material eventually obtained from tissues seems to behave as a rubber. Rubbers are materia s in which the work of deformation is largely recoverable and in which strain energy is stored in the form of changes of entropy rather than deformation of chemical bonds. In order that energy should be stored as changes in entropy, it is necessary that interactions between neighbouring chains should be low. In protein rubbers this condition is met only in the presence of water. A similar behaviour is observed in other polymer systems where substances of lower molecular weight are added to retain an easily deformable state. These substances are known as plasticisers and so by analogy water may be said to act as a plasticiser for elastin.

Mechanical properties of any polymeric material depend on temperature, frequency or strain rate, amount of plasticiser, the number of strong, permanent bonds between chains (usually termed cross-links) and the number of bulky side groups on the main carbon skeleton. At low temperatures, high frequencies, low concentration of plasticiser and high degree of cross-linking most rubbers become brittle glass-like substances. The transformation has been termed the glass–rubber transition. The exact conditions at the transition vary widely from polymer to polymer and so investigations normally use some form of oscillatory system in which stress, strain and time can be measured. The component of stress in phase with the strain is used to calculate the normal modulus and the out-of-phase component the storage modulus; the ratio between the two, which is the tangent of the lag angle, is a measure of the energy absorbed in each cycle. As the material, starting in the rubbery state, is cooled, the lag between stress and strain starts to increase and more energy is absorbed per cycle. Over a small range of temperature the normal modulus increases sharply and the energy loss in each cycle starts to decrease. This is the transition. The actual temperature is taken as the mid-point of the modulus change, which is normally about the temperature of the peak of the energy loss. As frequency increases, transitions occur at higher temperature. This can be expressed as:

$$\log a_T = c_1^0 (T - T_0)/c_2^0 + T - T_0$$

where a_T is the temperature shift factor, T_0 the reference temperature of the master curve, T the temperature of each experiment, c_1^0 and c_2^0 are empirical constants which apply at the reference temperature T_0. The equation allows modulus measurements over a restricted range of frequencies to be taken at different temperatures and shifted to form a composite curve covering a wider range of frequencies at a particular temperature.

The dynamic properties of elastin prepared by repeated autoclaving from ligamentum nuchae have been investigated by Gosline & French (1979). The elastin was first tested in distilled water at 36 °C, but because of resonances in the apparatus accurate measurements could not be made above 200 Hz and they were unable to reach the glass–rubber transition point. To their surprise, when the temperature was lowered to 2.5 °C similar behaviour was observed. The explanation lies in the increased absorption of water which occurs as the temperature is lowered (from 0.41 g water/g protein at 55 °C to 0.76 g water/g protein at 2.5 °C) which effectively plasticises the fibres. In order to retain the same degree of hydration, elastin must be tested in a non-aqueous environment.

Samples have been prepared by removing free water and equilibrating for 5 days over 0.05 M sodium chloride. Plunging the specimen into mineral oil retained the same degree of hydration when the temperature was lowered. The available range of frequencies at different temperatures permitted the production of master curves for 0.41 g water/g protein with a transition temperature of −15 °C and also at 0.46 g water/g protein with a transition temperature of $-26\,°C$. These compare with a transition of $10\,°C$ for 0.31 g water/g protein obtained by Kakivaya & Hoeve (1975) using scanning microcalorimetry. They were unable to use higher concentrations because below $0\,°C$ the free water freezes. The hydration of elastin is an important factor affecting the mechanical properties.

Native elastin fibres dissected from ligamentum nuchae have been studied by Carton, Dainauskas & Clarke (1962). In these samples no loss of the secondary components surrounding the core occurred during preparation, although the collagen fibrils surrounding the core could not be detected with a Masson Trichrome stain. The usual form of three-phase curve was obtained, but this time Phase II appeared at about $\lambda = 1.2$ and fracture occurred at about $\lambda = 1.3$. No creep was observed over a 2-minute period and strains up to about 90% of the rupture were recoverable. However, when Minns and his colleagues (1973) studied ligamentum nuchae in which the elastin had been isolated by formic acid treatment the breaking strain was much lower. Two explanations are possible: one is that the elastin had been damaged by the formic acid as suggested by Mukherjee (1969) and the other that the formic acid had removed the collagen fibrils spiralling around the elastic core and the elastin network had started to rupture at a weak point because the collagen was no longer available to give it support. Once rupture had commenced it would spread catastrophically through the network.

It is clear that the properties of elastin are extremely complex. One of the major difficulties is to distinguish between the elastic elements and their non-elastic components. The experiments of Carton *et al.* are among the few which deal with elastin fibres associated with variable amounts of additional components. Because of the unknown effects of these additional substances on the overall mechanical properties and the effects of chemical treatments on the tertiary structure of the protein, the relationship between the mechanical properties of chemically isolated elastin and the elastin in the natural fibre is unknown.

STRATUM CORNEUM

As the cells of the epidermis move away from the basal layer keratin is deposited and the cells flatten to become dead, horny squames eventually lost to the environment. This layer of dead corneocytes is the first recipient of all the mechanical forces applied to the body. Its mechanical properties, which are affected by temperature, hydration and chemicals, are therefore of great interest.

The properties of the corneum are governed by its constitutents which include keratin, various proteins associated with the cell membranes and a complex lipid and mucopolysaccharide containing intercellular material. Intercellular bonding which is mediated by desmosomes (specialised regions of the cell membrane), and interactions between the intercellular material and the cell membrane, greatly affect the mechanical properties of stratum corneum. At one time keratin was felt to be the component which determined the mechanical response, but since rupture of the corneum takes place along cellular junctions it is not the limiting factor and hair or wool are more representative of the mechanical properties of pure keratin.

Some mechanical tests have been carried out on stratum corneum still attached to the underlying dermis, but the majority have used isolated corneum. Chemical treatment is necessary to split the corneum from the dermis. With excised skin, exposure to ammonia vapour (Vinson *et al.*, 1964) or overnight incubation at 37 °C with the dermis in contact with 1% trypsin in 2 M urea at pH 7.2, using sodium bicarbonate to adjust the pH, will enable the corneum to be gently peeled away from the dermis (Park & Baddiel, 1972*a*). Stratum corneum may be obtained from volunteers by treating the skin with cantharidin-impregnated filter paper under occlusive patches for 4 hours and carefully protecting the resulting blisters until fully formed (Wildnauer *et al.*, 1971). Sunburn has been used by Elfbaum & Wolfram (1970) to liberate the stratum corneum.

Like most tissues stratum corneum is very sensitive to environment, especially relative humidity (r.h.). Humidity can be measured quite accurately using a wide range of meters or a wet and dry bulb thermometer, but both systems tend to be inaccurate near saturation and the identification of true 100% r.h. can be difficult. At saturation, condensation of water on the specimen becomes a complicating factor.

Some idea of the effects of humidity can be gained from the experiments of Park & Baddiel (1972*b*) using corneum from pig ear. Strips of corneum were equilibrated at high humidity, stretched at 10% and then allowed to dry under a very low load before final testing. This pre-treatment reduced the inter-sample variability. Specimens were then equilibrated at various humidities and mechanically tested. A short, initial, low modulus region gave way to an almost linear region and there was some evidence of a second low modulus region before fracture. The transition to this second low modulus region was termed 'yield'. The extensions at 'yield' varied from 1% at 30% r.h. to 20% at 100% r.h. The effective modulus in the linear region was from 2×10^9 N/m² at 30% r.h. to 6×10^6 N/m² at 100% r.h., a change in modulus of more than two orders of magnitude. No effort was made to correct for any changes in thickness with humidity.

The effect of small holes on rupture strain is unknown. Defects, especially of small radius, in brittle materials concentrate the stress locally and may initiate premature failure. Such small holes occur around the entries to sweat glands and hair follicles on isolated specimens. *In vivo*, catastrophic failures would be prevented by the underlying dermis and so, microscopic splits only would be observed. However, skin may undergo area extensions of between 50 and 100% and the arrangement of the tissue is probably of great significance. It has been suggested that it is the creases in young skin and

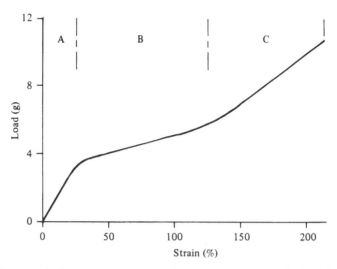

Fig. 38. A typical stress–strain curve for stratum corneum hydrated in water, illustrating the three behaviour regions. The tissue was pre-conditioned for 1 hour at 25 °C before application of load.

wrinkles in older tissue that allow sufficient movement to accommodate the difference in extensibility between epidermal cells and the stratum corneum (Ferguson & Barbenel, 1981).

However, much larger extensions have been reported, with variations from 20% at 0% r.h. to 190% at 100% r.h. for human stratum corneum when obtained from cantharidin blisters (Wildnauer *et al.*, 1971). Although their tests were carried out in air, Wildnauer *et al.* claimed completion within 10–30 seconds after removal from the humidity chamber. It is unlikely that changes in hydration could explain the differences recorded and a convincing explanation for the two sets of data has still to be found. There is the possibility that the data may be species dependent since Park & Baddiel used corneum from pigs' ears whereas Wildnauer *et al.* (1971) and Wilkes & Wildnauer (1973) used human tissue, although at the histological level the two tissues have similar appearance.

Tensile test curves for stratum corneum are qualitatively similar to those for whole skin, but show evidence of a second low modulus region following the high modulus portion of the curve (Park & Baddiel, 1972*a*). By contrast, Wildnauer and his colleagues (1971) obtained data showing much higher 'yield' and the initial low modulus region absent. In fact, if human corneum is pre-conditioned in water for 1 hour, a curve is obtained where two high modulus regions are separated by one of very low modulus (Fig. 38). In order to avoid confusion with the pattern for a normal stress–strain curve, the appropriate regions in Fig. 38 are labelled zones A, B and C.

Similar behaviour could be obtained when stratum corneum was equilib-

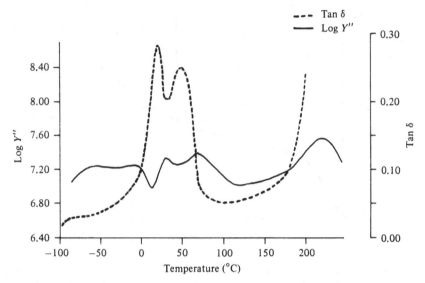

Fig. 39. Dynamic mechanical data as a function of temperature for a 56-year-old human; stratum corneum sample. Tan δ is the change in the viscous component and log Y'' is the out-of-phase component, where $Y'/Y'' = \tan \delta$. The viscous and modulus components exhibit independent changes with temperature.

rated in air at 98% r.h., but at 76% r.h. only the early part of the transition to zone B was observed while at 32% r.h. the material remained in zone A. As the humidity increases, the modulus in zone A and the ultimate tensile strength decrease while the work of fracture (measured as the area under the stress–strain curve) and the elongation at fracture increase.

Stratum corneum is also affected by temperature. Using an oscillatory testing system at frequencies of 100 Hz, 11 Hz and 3.5 Hz corneum was tested at temperatures from $-130\,°C$ to $250\,°C$ in a dry atmosphere (Wilkes & Wildnauer, 1973). The output curves are more complex than those obtained for elastin and at least three and possibly five transitions are seen (Fig. 39). Two definite transitions occur in the physiological range, one at about $0\,°C$ and the other at about $50\,°C$; the first is probably due to a change in the water binding and the second to a change in the lipid component. Above $220\,°C$ the specimen starts to discolour and so the apparent transition at about $200\,°C$ probably represents the onset of degradation rather than a true glass–rubber type transition.

If the corneum is first treated with chloroform and methanol, a procedure which greatly reduces the lipid component, then the peak at $45\,°C$ is much reduced or absent, but the peak near $0\,°C$ is unaffected. This adds support to the suggestion that the transition at $50\,°C$ is related to the lipid component. However, it should be noted that all these curves were obtained with the temperatures gradually increasing and they are not reversible. Since X-ray diffraction studies have shown that there may be an ordered lipid phase in stratum corneum (Wilkes, Nguyen & Wildnauer, 1973) it is possible that the 45 °C peak may involve an order–disorder transformation, and there is likely to be a much greater temperature lag on cooling than on heating. Since water acts as a plasticiser for elastin and the tests took place in dry air, there may be a continuous loss of water which would affect the transitions.

Other factors affecting the modulus include ionic concentration. If corneum is equilibrated in salt solutions there will be two effects: the effect of the ions, and change in thermodynamic activity of water which corresponds to an effective change in the relative humidity. The two effects may be distinguished by taking advantage of the fact that in a closed container the thermodynamic activity of the solvent is the same in the vapour and liquid phases and therefore if a piece of corneum is equilibrated above a salt solution in a closed container and another piece in the solution then the activity of the water in the two samples will be the same and differences will be due to the effects of the ions. In all the ionic solutions tested, the zone A modulus of the samples equilibrated in solution was lower than that of the samples equilibrated in vapour (Park & Baddiel, 1972c). It is suggested that the ions suppress interactions between opposite charges on the protein molecules making deformation easier. Sucrose reduces the activity of the water, but does not show any other effects on the modulus of intact corneum although it does reduce the modulus if the corneum is first extracted with solvents. However, tests by Van Duzee (1978) indicate that the modulus is a function of water content and that the effects of treatments such as immersion in salt solutions which produce no permanent effects may be to alter the water content of the corneum at a given water activity. Treatment with 3 M urea hardly affects the water content at 20% r.h., but at 90% r.h. the water content is reduced by a factor of 3. Similar results can be obtained with 3 M lithium bromide. When the measured modulus is plotted against water content instead of relative humidity or water activity, then all the results seem to lie on the same curve irrespective of treatment.

Changes in mechanical properties of stratum corneum are of interest when studying the response of skin to various chemicals. Water extraction alone does not alter the breaking strength although it does reduce the extension at fracture when tested at 100% humidity. Extraction in ether, or ether and then water, has little effect at low humidity, but at high humidity the breaking strength is greater, though the extension at fracture is less (Wildnauer *et al.*, 1971). If corneum is treated with chloroform or sodium dodecyl sulphate then the modulus of the treated corneum is higher than the untreated at all humidities (Park & Baddiel, 1972b).

The effects of chemical treatments may also be compared by the use of the Work Index, i.e. the work performed in stretching a treated sample by 5% compared with the work required to stretch an untreated sample in distilled water by the same amount. Control tests indicate that repeated tests in distilled water produce only insignificant changes (Elfbaum & Wolfram, 1970). Dimethyl sulphoxide (DMSO) produces no changes at concentrations below 50%, but at higher concentrations there is a reversible increase in the Work Index, probably due to swelling of the keratin. The Work Index is also increased reversibly by 5% trichloroacetic acid (TCA), with some reversibility by 2% and 5% phenol and irreversibly by 10% formaldehyde. Ammonium thioglycollate reduces the Work Index with partial reversibility. Since thioglycollates reduce cysteine bonds this provides some evidence for the mechanical role of disulphide bridges in stratum corneum. Protein denaturing agents (6 M urea and 97% formic acid), as expected, weaken the tissue. Some of the effects of urea can be reversed, but not those due to formic acid. It is thought that TCA may increase salt-like linkages, and that phenol increases hydrophobic bonding and may precipitate soluble proteins; formaldehyde is a well-known cross-linking agent.

It is possible to detect changes in stratum corneum that is still attached to the body (Christensen *et al.*, 1977). Using very low loads (1–10 g) and sinusoidal oscillations, differences in the modulus and hysteretic losses of normal and icthyotic corneum could be shown. Addition of small amounts of water to the corneum under test produced changes which reversed in about a minute, but similar changes produced by oil-in-water emulsions lasted over 3 hours.

Other techniques for the study of stratum corneum *in vivo* were presented by Nicholls & Marks (1977). One device, termed a 'cohesograph', measures the vertical force necessary to detach an 8-mm diameter disc glued onto the skin surface with cyanoacrylate adhesive. A positive correlation can be shown between an increase in the vertical force required and an increase in peak height as measured by surfometry. Another device, termed a 'scrubometer', uses a Perspex blade to scrub gently in a controlled manner a known area of skin bathed in a test solution. The detached corneocytes can then be collected, counted and examined in a microscope. It is suggested that this latter device tends to measure the bonding in the plane of the stratum corneum, i.e. shearing or tearing forces, while the cohesograph measures forces perpendicular to the plane. A combination of these techniques would allow the shedding of corneocytes to be studied, and an inverse relationship between the vertical detachment force and the number of squames released has been shown (King, Nicholls & Marks, 1981). An increase in the detachment force was shown in patients with scaling disorders, but there was a considerable overlap with the normal range and statistically significant differences could not be shown. The difficulties experienced by patients with scaling disorders show that retention of the corneocytes, especially if the corneum becomes dry and brittle, may be almost as troublesome as fragile corneum.

NOTATION

a	constant	R	gas constant
a_t	temperature shift factor	t	time
A	area	T	temperature
b	constant	T_0	reference temperature
c_1^0, c_1^0	constants applying to frequency	Y	elastic modulus in tension
	response curves at temperature T_0	ϵ	strain
d	thickness	$\bar{\epsilon}$	apparent strain
D	bending rigidity	$\dot{\epsilon}$	$d\epsilon/dt$: instantaneous creep rate
e	extension	θ_0	crimp angle
e_∞	$\sec \theta_0 - 1$	Λ	$4\epsilon/b^2 Q^2$ for collagen fibre where
ΔH	activation energy		$y = a \sin bx$
I	moment of inertia	σ	stress
l	length	ψ	$\dfrac{d}{l}\left[\dfrac{(1-\cos\frac{1}{2}\theta_0)}{\cos\theta_0}\right]$
Q	$(4\,I/A)^{\frac{1}{2}}$		

THERMAL PROPERTIES OF SKIN

The temperature of a local area of skin is governed by the external temperature, the internal body temperature, local blood flow and the metabolic rate of the tissues. Heat may be transferred from the area by conduction, convection and radiation and if these processes are insufficient to remove the total heat input, then the temperature will rise by an amount dependent on the thermal capacity of skin. In any homogeneous medium the equation for unidirectional heat flow may be written

$$q = -kA\,dT/dx$$

where q is the heat flow, k the thermal conductivity, A the area perpendicular to the direction of flow and T the temperature.

However, in all tissue, the problem is complicated by blood flow, which will remove heat very efficiently, and by the internal production of heat by metabolic processes. The more complex equation to describe a dermal layer is

$$\rho C \frac{\partial T}{\partial t} = k\nabla^2 T + W_b C_b(T_b - T) + Q_0$$

where ρ is the density, C the specific heat of tissue, W_b the blood flow, C_b the specific heat of blood, T_b the temperature of blood, Q_0 the energy generation from metabolic reactions (Cohen, 1977).

Measurements *in vivo* necessarily involve contributions from blood flow and changes in thermal conductivity have been used to indicate changes in blood flow. Several techniques have been used: thermocouples below various forms of contact heaters or infra-red beams, contact calorimeters or optical emission measurements. Thermocouples placed sub-cutaneously probably distort heat flow patterns, but surface contact may affect capillary blood flow and so determinations using only optical irradiation and emission have been attempted (Weaver & Stoll, 1969). Interpretation of optical experiments requires knowledge of the transmission, absorption and reflection of the different layers of skin at the appropriate wavelengths as well as the thermal emissivity.

Surface methods use some form of small localised heater and the temperature change is detected by a contact ring; this may be divided into separate segments to detect anisotropic heat flow. If metabolic heat production is negligible then at a steady state

$$0 = k \nabla^2 T + W_b C_b (T_b - T)$$

since $\rho C \partial T / \partial t = 0$ and $Q_0 = 0$.

In many cases an 'effective' thermal conductivity is obtained in which the contribution of blood flow is included in the constant

$$k_{\mathrm{eff}} \nabla^2 T = 0$$

For irradiation the change in temperature for an opaque solid has been derived as

$$\Delta T = \frac{2 \mathcal{A} \, \mathcal{H} \, t^{\frac{1}{2}}}{(\pi \, k \, \rho \, C)^{\frac{1}{2}}}$$

and for a diathermous solid

$$\Delta T = \frac{\mathcal{A} \, \mathcal{H}}{k} \left[2 \left(\frac{ht}{\pi} \right)^{\frac{1}{2}} + \frac{\exp(\lambda^2 \, ht)}{\gamma} \operatorname{erfc} \gamma \, (ht)^{\frac{1}{2}} - \frac{1}{\gamma} \right]$$

where ΔT is the temperature rise, $\mathcal{A}$ the radiant absorbance, $\mathcal{H}$ the irradiance in cal cm^{-2} s^{-1}, k the specific thermal conductivity in cal cm^{-1} °C^{-1}, ρ the density in g cm^{-2}, C the specific heat in cal g^{-1} °C^{-1}, h the thermal diffusivity $(k/\rho C)$, γ the optical extinction coefficient in cm^{-1} and

$$\operatorname{erfc} \gamma = 1 - \frac{2}{\pi^{\frac{1}{2}}} \int_0^y \exp(-x^2) \, \mathrm{d}x$$

The product $k\rho C$ may be measured from the surface temperature of blackened skin by use of the equation

$$k\rho C = \frac{4 \mathcal{A}^2 \, \mathcal{H}^2 \, t}{\pi (\Delta T)^2} = \frac{1.15 \, \mathcal{H} \, t}{(\Delta T)^2}$$

since $\mathcal{A} = 0.95$ for blackened skin (Derksen, Murtha & Monahan, 1957).

A review of thermal conductivity measurements has been undertaken by Cohen (1977) and some relevant values are given in Table 8.

Thermal constants should be measured by methods appropriate to their use. For example Buettner (1936) and Weaver & Stoll (1969) have shown that the effective thermal conductivity of skin can increase by an order of magnitude when the skin is warmed. This increase is a measure of vasodilation which occurs in a time scale of seconds. If damage due to heating regimes in which the temperature rise occurs relatively slowly were being investigated then the higher values of conductivity would be appropriate. If flash burns where intense heat is applied for a very short time were being studied then lower values must be used.

TABLE 8. *Thermal conductivity of skin*

Tissue	Thermal conductivity (cal cm^{-2} s^{-1} °C^{-1} × 10^{-4})	Reference
Human epidermis	5.0	Lefevre (1901)
Porcine epidermis	5.0	Henriques & Moritz (1947)
Human dermis	7.0	Roeder (1934)
Skin *in vitro*	7.7	Lipkin & Hardy (1954)
Upper 2 mm	9.0	Buettner (1936)
Cool	13.0	Buettner (1936)
Warm	67	Buettner (1936)
Blackened	8–75	Weaver & Stoll (1969)
Normal	9.0	Lipkin & Hardy (1954)

NOTATION

A	area	T	temperature
$\mathscr{A}$	absorbance	T_b	temperature of blood
C	specific heat	x	spatial coordinate
C_b	specific heat of blood	γ	optical extinction coefficient
h	thermal diffusivity	Δ	increment
$\mathscr{H}$	irradiance	λ	wavelength
k	thermal conductivity	ρ	density
q	unidirectional heat flow	∇^2	$\frac{\partial^2}{\partial x^2}+\frac{\partial^2}{\partial y^2}+\frac{\partial^2}{\partial z^2}$
Q_0	energy generation due to metabolic reactions	erfc	error function
t	time		

ELECTRICAL PROPERTIES OF SKIN

Skin has a set of complex electrical properties. The most commonly used method of description involves equivalent circuits while the alternative method attempts to model skin in terms of the theories of solid state physics. Factors such as site, mechanical deformation, temperature, relative humidity and the state of arousal of the subject as well as injury or pathological conditions are all known to affect the results and so precise values for the components of the equivalent circuits or coefficients of the mathematical equations are difficult to establish.

In most studies, some form of electrode is placed on the skin. A standard electrode has an area of about 1 cm², but micro-electrodes with areas of a few square micrometres are also used. These are assumed to sample a much smaller volume of tissue since site to site variation over very small distances is common. Also, the distance from an active anatomical site, for example a sweat gland, has a marked effect on the values recorded.

Skin is usually prepared by washing with soap and water or swabbing with solvent to remove surface grease and loose debris. After this pre-treatment there will still be a potential measurable on the surface of the skin which will change by about 5 mV if the skin is deformed. This surface potential can be troublesome in certain types of measurements, especially electrocardiograms or electroencephalograms, and various techniques for removing it have been tried. These most commonly involve removal of the stratum corneum by abrasion, which can lead to uncomfortable skin reactions if an irritant electrode paste or jelly is used to improve contact. A controlled micro-puncture technique has been advocated by Burbank & Webster (1978) who also used a non-irritating electrode cream composed of 0.7 ml of saturated sodium chloride solution in a 1.5 oz tube of Johnson and Johnson First Aid Cream to give approximately 0.9% sodium chloride by weight. As a particular factor such as site or humidity is varied the changes in values are often of more interest than the absolute value obtained. Even then, the effects of extraneous variables must be removed wherever possible because these may be large enough to obscure the real relationship being studied.

Stratum corneum has a higher resistance than the interior of the body. Most current flow in the body takes place by diffusion of ions and this includes ionic diffusion through the layers of the stratum corneum. In special circumstances, such as dry stratum corneum which inhibits ionic diffusion, various forms of solid state charge transfers may occur. Resistance to current flow between surface electrodes depends both on the resistance of the stratum corneum and on the existence of low-resistance paths to the interior of the body. Sweat glands and hair follicles seem to offer such paths because of their relatively high water content.

The relationship between skin resistance and the number of active sweat glands per unit area has been difficult to establish since the electrodes themselves prevent evaporation of sweat and so cause changes in the hydration of the stratum corneum. Water moves from the interior of the body to the external environment and the flux of water, J_s, can be written as

$$J_s = [D(N_1 - N_2)]/d$$

where D is the diffusion coefficient, N_1 and N_2 are the water concentrations at the epidermal boundary and outer surface respectively and d is the thickness of the stratum corneum. If an electrode is used which prevents evaporation, then the water content of the stratum corneum rises until eventually N_2 equals N_1. The time for this process is about 16 minutes (Campbell *et al.*, 1977). When electrodes are to be left in place for any length of time, precautions must be taken to minimise alterations in hydration.

A positive correlation exists between skin conductance (the reciprocal of resistance) and the number of active sweat glands when using brief contact with dry metal electrodes (Thomas & Korr, 1957). However, if the skin is occluded, then correlation coefficients are lower and usually not significant.

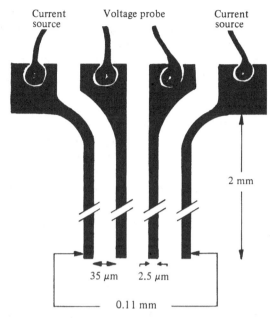

Current source   Voltage probe   Current source

2 mm

35 μm   2.5 μm

0.11 mm

Fig. 40. A four-line micro-electrode probe.

Electrode solutions, jellies or pastes which alter the hydration of stratum corneum also obscure this correlation. Micro-electrodes have been used to demonstrate that a filled sweat duct has much lower resistance than the stratum corneum between glands (Edelberg, 1977). If the number of active glands under a standard electrode is counted and the resistance calculated by using an array of parallel resistors connecting each electrode with the interior of the body, this resistance is about twice as high as the resistance measured using isotonic saline as the electrode solution, indicating that wet corneum has an appreciable conductivity.

Stratum corneum which had been treated in various ways to alter water binding properties was studied by Campbell *et al.* (1977) using a four-line micro-electrode system (Fig. 40). In such a system, assuming infinite thickness, conductivity can be expressed as

$$\rho = \frac{\pi l}{2 \ln 2} \frac{V}{I}$$

where ρ is the conductivity, V the voltage measured through the inner electrodes, I the current injected through the outer electrodes and l the electrode length. The equation is accurate to 2% if the material studied has a thickness 10 times the electrode spacing and the electrode length is large compared with the electrode spacing. With an electrode spacing of 35 μm, Campbell's assembly was used to study excised plantar corneum. The resistivity of untreated corneum was found to vary over three to four orders of magnitude as hydration was changed. Treatment with formaldehyde increased conductivity for a given water content while treatment with urea reduced it. Formaldehyde is believed to reduce water binding and the increased amount of free water enhances ionic diffusion and so lowers resistivity. Urea, on the other hand, increases water binding and thus reduces conductivity.

Suitable micro-electrodes may also be used to investigate changes in resistivity through skin. As an electrode contained within a micro-pipette is slowly pushed through the stratum corneum, resistance falls. This might be expected from the model describing the diffusion of water through the stratum corneum since this assumes that under normal conditions hydration increases with depth. In practice there is a gradual fall followed by a sudden drop in resistance at about 50 μm on the forearm and 350 μm on the palm. Interpreting this phenomenon, Edelberg (1971) suggests that there is a discrete barrier to ionic permeability in addition to the continuous one of the corneum; this discrete barrier may be the stratum granulosum.

The effects of changes in ionic concentration can be demonstrated in the foot pad of an anaesthetised cat by soaking the pad in de-ionised water, isotonic or hypertonic saline, then dehydrating the pad in dry air. Rehydration can be carried out by stimulating the sweat gland nerves (8 V d.c.; 2-ms pulses; 3 Hz). The conductance at any given hydration varies with ionic concentration (Steinmetz & Adams, 1981).

High temperature and humidity will tend to activate sweat glands, which hydrate the stratum corneum, and so resistance will be low. If the humidity is low, then the corneum will be dry and conductance will be through the sweat glands and hair follicles. Psychological stress or pathological conditions which increase sweating will also tend to lower resistance. Conversely disorders which inhibit sweating will raise resistance, while areas of the body which have a plentiful supply of hair follicles or sweat glands will tend to have relatively low resistance.

In dry corneum other mechanisms for charge transfer may become operative. Many biopolymers in the dry state have been shown to possess intrinsic semiconduction where charge is carried by electrons or holes produced by thermal excitation across a small energy gap between the conduction and valence bands (Pethig, 1979). However, in the presence of bound water, conduction may occur by electron tunnelling between water molecules. As the amount of water increases, ionic conductivity becomes more important until at a level of about 25% water, most of the conduction is ionic.

In general, biopolymers studied in the laboratory seem to obey the equation

$$\rho = \rho_0 \exp(-\xi/kT)$$

where ρ is the conductivity, ρ_0 is a material constant, k is Boltzmann's

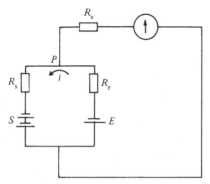

Fig. 41. Circuit for internal currents in skin. E, potential of epidermal generator; P, potential measured at surface; R_e, resistance of epidermis other than sweat duct; R_s, resistance of sweat duct and internal resistance of the sweat gland 'battery'; R_v, resistance of voltmeter; S, potential of sweat gland generator. The value of P is dependent upon the ratio of $R_s : R_e$. (From Edelberg, 1968.)

constant, T is the temperature in K and ξ is the activation energy. Collagen, elastin and keratin have values for ξ in the range 1.3–1.5 eV, ρ_0 in the range 10^6–10^7 and the conductivity at 300 K in the range 10^{16}–10^{18} ohm·m.

The natural electrical potential of skin varies from site to site over the body surface and is generally more negative than the interior of the body. Local changes in potential can only exist on the surface of materials of relatively high resistance. Typical values of surface potential would be -10 to -60 mV on the palm with the mean at about -35 mV when measured with reference to the interior of the body. In general the right hand has a greater negative potential than the left. A mean value of about -15 mV is found on the forearm (Edelberg, 1971).

Removal of the stratum corneum reduces the skin potential to very nearly that of the interior, emphasising the contribution of stratum corneum to skin resistance. The interior of the sweat duct through the epidermis is highly negative, but as the duct fills the resistance falls and the surface negativity also falls. The epidermis itself can generate a potential and the recorded voltage will be a function of the two main sources outlined above. An equivalent circuit was proposed by Edelberg (1968) (Fig. 41).

While the source of epidermal potential is still disputed, a popular theory based on selective ionic diffusion includes the idea that fixed charges on the outside of membranes reduce the effective mobility of negative ions. Some evidence for this theory comes from the relative diffusion coefficients of potassium (K^+) ions and chloride (Cl^-) ions in aqueous solution. The effective diameters of the hydrated ions are similar although the calculated diameter of the K^+ ion is much smaller than that of Cl^-. However, K^+ ions diffuse inwards much more rapidly than Cl^- ions, indicating that the diffusion coefficient is not governed simply by the diameter of the hydrated ion.

Therefore, if all negative ions have smaller diffusion coefficients then the outside of the skin would be expected to have a more negative potential. Unfortunately, the results of measurements on other electrolytes do not support this theory (Edelberg, 1977). Nevertheless, in experiments on cat foot pad, those pads with the highest ionic content also possessed the highest potentials (Steinmetz & Adams, 1981).

In living cells, the cell membrane potential induced by the sodium pump contributes directly to the tissue potential, but in stratum corneum, largely composed of dead cells, explanations for the maintenance of surface potentials are more difficult to find. Changes in resistance of sweat glands which short the potential can make detailed interpretation even more difficult.

Skin potential is also related to skin stretch; typically the changes are about 5 mV (Burbank & Webster, 1978). This can interfere with electrocardiogram recording from moving subjects. Careful siting of electrodes can reduce the effects, but micro-punctures minimise the effects by short-circuiting the skin resistance and permit lower noise electrocardiogram recording from moving subjects. The time scale over which these potentials change as a result of skin stretch would indicate some other mechanism for the production of potential rather than differential diffusion.

So far hydrated skin has been treated as through it was a simple ohmic resistor with some potential generators, but in fact it has capacitive elements as well. The resistance of skin to alternating currents therefore depends on frequency. To express this capacitive factor a number of equivalent circuits have been proposed and examples are shown in Fig. 42. These circuits do not assume that skin contains elements which have precise values of resistance and capacitance, but rather that the circuit will, in general, mimic some or all of the responses of skin to applied current.

If a material with a high resistance is placed between two parallel conducting electrodes and a potential applied to the electrodes the arrangement is termed a parallel-plate capacitor. The potential on the plates will attract charges within the insulating material of opposite sign to the potential and so increase the charge on the electrodes. There are three and sometimes four sources for the charge within an insulator. If the potential is high enough then the electron orbits around individual atoms may be distorted; this is termed electronic polarisation. If there are molecules within the solid, then there may be distortion of atoms in such a way that parts of the molecule become charged with opposite charges so that the molecule as a whole remains electrically neutral. Movement of these charges is termed atomic polarisation. However, some molecules have permanent charges, usually referred to as a dipole moment since in a neutral molecule the fractional charges of one sign must be counterbalanced by fractional charges of the other sign. An example of a molecule having a permanent dipole moment would be water. The oxygen atom carries a slight negative charge and each hydrogen atom a slight positive one. In an applied electric field the oxygen atoms would tend to orientate

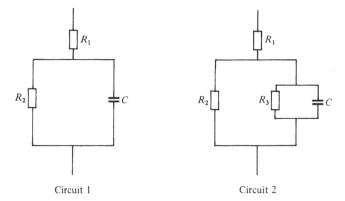

Circuit 1   Circuit 2

Fig. 42. Equivalent circuits for skin impedance. R, resistive elements; C, capacitive elements.

themselves towards the positive electrode and the hydrogen atoms towards the negative electrode. Thus, the total ability of a molecule to polarise, which alters the charges on a particular pair of electrodes, is the sum of the distortion of electron orbits around individual nuclei, the displacement of atomic nuclei with the consequent production of a temporary dipole moment and reorientation of any permanent dipole moment. In ionic solids, displacement of the individual ions may give rise to ionic polarisation. Some complex ions may also have permanent dipoles, but in the solid state these substances are unlikely to be completely ionised.

If an alternating potential is applied, rearrangements of the molecular dipoles permit a net transfer of charge between one plate and the other, i.e. a current may be passed which is greater than that which would be passed for a static potential. Secondly, rearrangements will take a finite time to accomplish and therefore the magnitude of the polarisation will vary with frequency. At a sufficiently high frequency the material will no longer respond by reorientation and so the polarisation will diminish. During the frequency range in which reorientation becomes difficult, the polarisation will lag behind the applied potential and so the dielectric constant is written as a complex number in which the real component represents the polarisation in phase with the applied voltage and the imaginary component the polarisation out of phase with the applied voltage.

Each orientation process can be characterised by a constant termed the relaxation time, τ, and the permittivity may then be written

$$\epsilon^* = \epsilon_\infty + (\epsilon_0 - \epsilon_\infty)/(1 + \mathrm{j}\omega t)]$$

where ϵ_∞ is the permittivity at a frequency sufficient to reduce the orientational polarisability to negligible proportions, ϵ_0 is the limiting low-frequency permittivity, ω is the angular frequency which equals $2\pi f$ where f is the applied frequency in hertz, and j is $\sqrt{-1}$. The ratio of the permittivity of the solid to the permittivity of a vacuum is termed the dielectric constant, K.

In a complex molecule there may be several relaxation processes and this is especially common where a molecule has several different side chains; so polypeptides would be expected to have complex relaxation spectra. For a single relaxation the plot of ϵ', the in-phase component, against ϵ'', the out-of-phase component, will be a semicircle; this is often referred to as the Cole–Cole arc after the authors of the original paper (Cole & Cole, 1941). Unfortunately, many equivalent circuits in which a plot of in-phase and out-of-phase currents forms some sort of arc are described as following the Cole–Cole arc; the mathematics may be similar, but the physics is not and the mechanisms which produce the plot may be quite different.

In a material such as skin a study of the response to alternating current will be difficult to interpret. There are many different complex molecules present and these are surrounded by water molecules associated with varying numbers of ions. In general, reorientation of dipoles in response to an applied voltage is easier and quicker in solution than in the solid state and so measurement of relaxation times in the solid is only a very rough guide to times in solution. If ions are dissociated then complete charge separation may occur and free hydrogen has been detected at electrodes.

In order to simplify an obviously complex situation various models have been proposed. The simplest models select a particular arrangement of capacitor and resistors which mimics the response at a particular frequency. Two typical circuits are shown in Fig. 42. It is necessary to use more than one resistor because the impedance of skin at very high frequency is low, but not zero, while the low-frequency or d.c. impedance is very much higher and at these frequencies capacitors pass very little current. Considering the equivalent circuit 1 (Fig. 42), the d.c. resistance is equal to $R_1 + R_2$. At very high frequency the reactance of the capacitor becomes negligible and so the impedance becomes equal to R_1. The values of the reactance of human skin do not follow a simple model and so the response has to be matched against a chosen equivalent circuit which allows values for the components to be chosen to achieve a close fit to the measured values. A modified circuit to fit certain conditions is shown as circuit 2 in Fig. 42. But, even when the component values of an equivalent circuit can be estimated, these only represent 'lumped' parameters. Electrical pathways through skin are many and do not behave as simple resistors and capacitors. A better description of skin behaviour is obtained by studying the response at many frequencies.

One of the difficulties in studying living skin is to isolate its response from that of the underlying tissues. Perhaps the best way is to use a closely spaced multi-electrode system which reduces the effective sample volume. The four-line micro-electrode system used by Campbell *et al.* (1977) would be an example (see Fig. 40). The values of the components in the equivalent circuits are all frequency dependent with the exception of R_1.

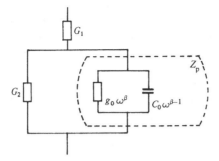

Fig. 43. Equivalent circuit of skin impedance. $G = 1/R$ and $g = \delta I/\delta V$.

A simple circuit was proposed by Yamamoto *et al.* (1978) in which the frequency response was included (Fig. 43). The values of the components are dependent on current density and in non-linear circuit elements a differential component is used to define the different parameters. Lower-case letters are normally used to distinguish the differential form from the linear relationships in general use. Conductance of a non-linear element, g, is $\delta I/\delta V$ and this should be contrasted with the conductance of a linear element, G, which is I/V.

The values of the components in the circuit can be adjusted to give a semicircular arc when the in-phase and out-of-phase components are plotted if the following relationships are assumed. If the phase angle ϕ is a constant then ϕ may be written as $\phi = \beta\pi/2$ and the polarisation impedance Z_p of the circuit is then

$$Z_p = af^{-\beta}\exp[-\mathrm{j}\beta\pi/2]$$

and the total impedance

$$Z - R_\infty = \frac{1}{1/R_2 + 1/Z_p} = \frac{R_2}{1 + (\mathrm{j}f/f_m)^\beta}$$

where a is a constant and f_m is $(a/R_2)^{1/\beta}$. Thus this circuit can be adjusted to mimic the known frequency response of skin (Cole, 1940; Yamamoto & Yamamoto, 1976).

Although the equivalent circuit approach allows changes in electrical response of skin to be compared, it does not define the anatomical structures producing the response. A variety of theoretical models have been proposed which aim to predict skin response using information on the properties of different tissue components and their organisation. For example, the aqueous channels in the stratum corneum and epidermis are believed to form the analogues of the capacitor 'plates'. This enables the size of the capacitive elements to be estimated.

For a parallel-plate condensor the capacitance can be written

$$C = (0.089\,KA/d) \times 10^{-6}$$

where C is measured in microfarads (μF), K is the dielectric constant, A is the area in square centimetres and d the thickness in centimetres (Edelberg, 1977). The capacitance of skin is in the range 0.02–0.06 μF, and so if the dielectric constant is 10 and the capacitance 0.02 μF/cm², the thickness of the capacitors would be about 0.5 μm and, if K is 5 with a capacitance of 0.05 μF/cm², about 0.1 μm. Both values are considerably thicker than a cell membrane but smaller than a cell. It is likely, therefore, that the capacitance is due to a more complex form of polarisation than the analogue of a simple insulator between two parallel aqueous channels.

So far, various models have successfully predicted the correct phase angle of current to voltage, but the plot of the real and complex portions of the impedance as a function of frequency does not give rise to the experimentally observed semicircular plot. The mathematical equation of such an impedance, Z, is

$$Z = R_s + j X_s = R_\infty + \frac{R_0 - R_\infty}{1 + (j\omega\tau_p)^{1-m}} \quad (0 < m < 1)$$

where R_s is the 'lumped' resistance, X_s the 'lumped' reactance, R_∞ is the value of R_s at very high frequency, R_0 the value of R_s at very low frequency, τ_p a mean relaxation time and m is related to the width of the relaxation spectrum. The analogy with the equation for a dielectric is obvious, but the origin of the response is different.

None of the more complex electrical models for skin, where specific structures are assigned particular electrical properties, produces such a plot unless some implausible assumptions about the variation of particular parameters are made. In the model proposed by Barnett (1938) and Tregear (1966) in which the resistance varies inversely with distance from the surface while the permittivity remains constant, the phase angle may be made similar to skin. The components of the Z locus are given by

$$\frac{R_s - R_\infty}{R_0 - R_\infty} = \frac{\text{arc tan}(\omega\tau_2) - \text{arc tan}(\omega\tau_1)}{\omega(\tau_2 - \tau_1)}$$

and

$$\frac{-X_s}{R_0 - R_\infty} = \frac{1}{2\omega(\tau_2 - \tau_1)} \cdot \ln\left[\frac{1 + (\omega\tau_2)^2}{1 + (\omega\tau_1)^2}\right]$$

where $\tau_1 = CR_{min}$ and $\tau_2 = CR_{max}$. This is quite unlike the Cole–Cole locus (Salter, 1981a).

A model using the theories of solid state physics assumes that, in the arrangements of proteins found in skin, a type of semiconduction may be produced. If the energy gap between conduction and valence bands is small then a few electrons may be thermally excited across the gap and the material is said to be an intrinsic semiconductor. But if certain types of additional material are present then a few electrons may be excited to energy levels between the two bands where they persist for a short period, during which charge may be carried; this is termed extrinsic semiconduction. These states of intermediate energy are known as localised states and the average period spent by a particular electron in that excited state is known as the relation time. The effect of these finite relaxation times is to give an effective capacitance, which is also seen in transistors. Salter gives the expression for the complex permittivity of such a material as

$$\frac{\epsilon^* - \epsilon_\infty}{\epsilon_0 - \epsilon_\infty} = \frac{(1-\alpha)\pi}{\sin \alpha \pi}, \quad \omega_{min} \lesssim \omega \lesssim \omega_{max}, \ 0 < \alpha < 1$$

where ω_{max} tends to infinity and ω_{min} depends on τ_2 and α and can be calculated.

If a d.c. shunt conductivity is present and ϵ_∞ is small then

$$\rho^* = j\omega\epsilon^* = \rho_0 + \frac{\epsilon_0(1-\alpha)\pi}{\tau_2 \sin(\alpha\pi)}(j\omega t)^\alpha, \quad \omega_{min} \lesssim \omega \lesssim \infty, \ 0 < \alpha < 1$$

and

$$Z = \frac{R_0}{1 + (j\omega\tau_p)^{1-m}} \quad 0 < m = (1-\alpha) < 1$$

if $(\omega\tau_p) > (\omega\tau_p)_{min}$ where

$$(\omega\tau_p)_{min} = \tau_2 \omega_{min}\left[\frac{R_0 \epsilon_0(1-\alpha)\pi}{\tau_2 \sin(\alpha\pi)}\right]^{1/\alpha}$$

In fact this expression for Z holds down to very low frequencies. Thus, this model will necessarily obtain the semicircular locus because of the physics used in its derivation. More details of the interpretation of this model can be found in Salter's thesis (1981*b*).

CHANGES DUE TO PATHOLOGICAL CONDITIONS

Attempts have been made to use differences in skin potential as an aid to the diagnosis of basal cell carcinoma and benign inflammatory lesions (Woodrough, Canti & Watson, 1975). All the lesions were on the face and the reference electrode was placed under the tongue. An assumption was made that contralateral normal sites would have similar potentials and the potentials of various non-ulcerated lesions were compared with the potential of the normal contralateral site. Benign inflammatory lesions have similar potentials to normal skin while basal cell carcinoma is considerably more positive. Unfortunately while the results for the carcinoma group show significant differences from the benign group, the populations overlap to such an extent that the technique cannot be used as a single diagnostic test.

Chemical damage also alters electrical properties. Treatment with sodium hydroxide at pH 12 causes a marked decrease in skin impedance. Solutions of pH 10 produce only a small decrease in impedance and subsequent treatment of the same site on a daily basis produces still smaller changes, indicating that skin may become habituated to mild chemical insult. The more severe insult given by solutions at pH 12 does not seem to produce any habituation (Malten & Thiele, 1973).

Dishydrotic eczema seems to involve hyperhidrosis of the soles and palms in some patients. Whether this hyperhidrosis, which may be emotional in origin or an allergic response, is the cause of the condition is still disputed. Miller & Coger (1979) decided to investigate the use of biofeedback techniques to reduce the hyperhidrosis and improve the condition. Two groups of patients were used; in one group the monitoring device was adjusted so that patients were encouraged to increase conductivity, so tending to increase water content, while the second group had the device adjusted so that water content was decreased. The first group of 12 patients reported difficulty in controlling their devices and 2 patients were withdrawn from the study after severe exacerbation of the condition. In the second group of 22 patients only one dropped out and 16 out of 22 improved, compared with 4 out of 12 in the first group. It seems that reduction of water content may improve dishydrotic eczema, but the results are difficult to interpret because of the placebo effect of participating in a clinical trial.

EFFECTS OF DRUGS AND COSMETICS

Many cosmetic products are described as 'moisturising' and methods, including electrical, of detecting changes due to their application have been sought. The effects of oil-in-water and water-in-oil emulsions in low-humidity environments were studied by Clar, Her & Sturelle (1975). Subjects sat with their forearms in a glove box held at 86% relative humidity (r.h.) and 25 °C, and after a 20-minute equilibration time the impedance was measured at 25 Hz. The humidity was lowered to 66% r.h. and after equilibration the measurements repeated. On subsequent visits an emulsion was applied to one forearm before repeating the tests.

In other experiments the lower humidity provided the reference state and sodium lactate and liquid paraffin were applied before raising the humidity. The sodium salt of pyrrolidone carboxylic acid (PCNa) may be added to preparations to increase the moisturising ability. Aqueous solutions containing 2% and 10% PCNa and a cosmetic preparation with and without 2% PCNa were applied to four separate sites and the changes compared with an untreated site. The last test compared readings before and after 14 days' use of a cosmetic containing 1% placental extract.

Results confirmed that impedance values were lower at 86% r.h. than 66% r.h. Interpretation of the changes due to treatment was difficult because penetration of the corneum by resistive oils altered the impedance; evidence of increased moisture content could be obtained by studying the impedance changes at a range of frequencies and by considering the change in relaxation time. Oil-and-water emulsions had little protective effect at lower humidities, unlike water-in-oil emulsions. The role of the base can be seen in the experiments on PCNa. Active agents such as PCNa and lactate increase water content, but the 10% aqueous solution only produced the same changes as the 2% cosmetic preparation. The cosmetic without PCNa produced no

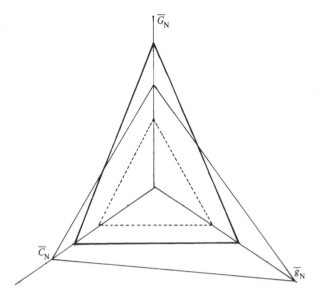

Fig. 44. Rate-of-change triangles for DL-2-pyrrolidone-5-carboxylic acid. The sodium salt is a natural moisturising agent present in human skin.

change. Even after 14 days of regular use placental extract had no significant effect.

A frequency response spectrum was obtained by Yamamoto *et al.* (1978) and used to evaluate the response in terms of the equivalent circuit shown in Fig. 43. By means of matrix transformations three parameters were obtained, one relating to the change in dielectric constant, one to conduction through a polarisable material and the last to ionic conductivity. These parameters were plotted on three-dimensional Cartesian axes and a triangle drawn connecting the parameters calculated from a particular measurement (Fig. 44). By plotting triangles pre- and post-treatment on the same axes, changes due to treatment were clearly displayed.

Oil-in-water emulsions decreased ionic conductivity 30 minutes after application but the effect was lost in 3 hours, while water-in-oil emulsions produced similar initial effects but these were subsequently sustained. A branched ester oil reduced conductivity about twice as much as the emulsions while PCNa increased the conductivity progressively over the 3 hours. It is possible, therefore, that the drop in conductivity is due to the penetration of the oil into the corneum. Similar results were obtained for the polarisation conductivity except that PCNa produced a very large increase at 30 minutes but had decreased at 3 hours, although the level was still higher than the initial value. The dielectric constant first decreased over a period of 30 minutes for the oil-in-water cream, but increased after 3 hours. It decreased and remained constant with the water-in-oil cream. Branched ester oil decreased the dielectric constant which then remained steady over 4 hours.

It seems that electrical changes where both oils and water are present in the same preparation are difficult to interpret. Assuming that PCNa increases moisture content and that the branched ester oil does not affect water content then the emulsions may simply reflect a decrease in all parameters. Oil-in-water emulsions merely have briefer effects.

Although the mathematics required is complex, changes are shown very clearly. The physical basis of the method has not been well explained and there is sometimes a confusion between the Cole–Cole arc for a dielectric and the change in impedance with frequency for skin. However, it appears that the general approach is worthy of further development and this method of presentation allows easy comparison between treatments.

EFFECTS OF PSYCHOLOGICAL STIMULATION

Psychological arousal, particularly anxiety, alters skin potential and skin conductivity, a response which has been used in 'lie detectors'. Skin potentials measured at rest appear to have a normal distribution with the central value for palm at about -35 mV with respect to the forearm (Surwillo, 1969), but if subjects are asked to perform a simple vigilance task the statistical distribution of potential in the young becomes skewed. Accuracy and vigilance did not differ between the young (under 53 years) and old (54 years and over), which is an indication that this response may not be simple.

Skin conductivity does not seem to be significantly affected by cognitive tasks such as simple mental arithmetic or multiple-trial free recall, even though heart rate is noticeably altered (Furchtgott & Busemeyer, 1979). Skin conductance does change, however, during simple reaction-time tasks where the level of financial reward depends on fast times. The amount of change increases as the number of tests on any particular subject increases. The latency between a warning light indicating the start of a test and the change in conductivity varied and subjects having longer latencies in general had poorer reaction times (Cowles, 1973).

More anxiety can be engendered by the use of a noxious stimulus such as a mild electric shock. Changes in skin conductivity have been investigated by Colgan (1970); on average, 50% of tests involved shocks. A slight decrease in conductivity occurred as the tests progressed, but in a group where, after 15 random tests, the subjects were instructed on how to predict shocks the conductivity change decreased markedly for subjects in 'non-shocked' tests, but remained high in those trials where shocks were given.

No evidence was found to support the view that the predictability of the noxious stimulus decreased anxiety.

BASIC DEFINITIONS

Resistance: The ratio between voltage and current. For alternating current the current and voltage are in phase. $V = IR$.

Capacitance: The ratio of voltage to stored charge (Q). $V = QC$.

Inductance: The ratio of induced voltage to changing magnetic field. This changing field is usually produced by varying the current flowing round a coil: thus $V = L\,dI/dt$.

Angular frequency, ω: This is equal to $2\pi f$ where f is the frequency in hertz. Using the angular frequency and the symbol j for the imaginary part of a complex number, the equations for the elements may be written as

$$V = V_0 \cos \omega t$$
$$I = I_0 \cos \omega t$$
$$V = IR$$
$$V = j\omega L I$$
$$V = I/j\omega C$$

When dealing with alternating currents and circuits involving components other than pure resistors, then some other terms are used:

Impedance: The ratio between an alternating voltage and current. $V = IZ$. If it is desired to consider the in-phase and out-of-phase components separately then $Z = R + jX$.

Reactance: The out-of-phase component of impedance.

It may be convenient in many cases to use inverse functions:

Admittance: $Y = I/V = 1/Z$ may also be written as $Y = G + jB = (R + jX)^{-1}$.

Conductance: G is the real part of the admittance. It equals $1/R$ for a simple resistance.

Susceptance: B is the complex or out-of-phase part of the admittance.

Resistivity: A material property that equals the resistance of 1 m of the material with a cross-sectional area of 1 m².

Conductivity: The reciprocal of resistivity. In materials in which do not form simple resistors conductivity may be complex (reciprocal of the impedance of 1 m of the material with a cross-sectional area of 1 m²) and then:

Permittivity: Q/E, where E is electric field.

Dielectric constant: $K = \epsilon_{MAT}/\epsilon_{VAC}$ and in practice represents the ratio of the capacitance of a parallel-plate condensor when the gap between the plates is filled with the material and when the gap contains a vacuum, thus $K = C_{MAT}/C_{VAC}$.

For resistances in series:

$$R = R_1 + R_2 + \ldots + R_n \quad 1/G = 1/G_1 + 1/G_2 + \ldots + 1/G_n$$

For resistances in parallel:

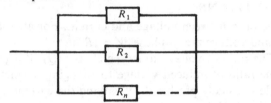

$$1/R = 1/R_1 + 1/R_2 + \ldots + 1/R_n \quad G = G_1 + G_2 + \ldots + G_n$$

For impedances in series:

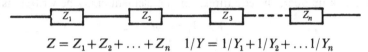

$$Z = Z_1 + Z_2 + \ldots + Z_n \quad 1/Y = 1/Y_1 + 1/Y_2 + \ldots 1/Y_n$$

For impedances in parallel:

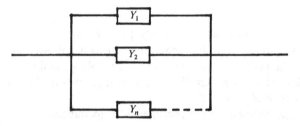

$$1/Z = 1/Z_1 + 1/Z_2 + \ldots + 1/Z_n \quad Y = Y_1 + Y_2 + \ldots + Y_N$$

NOTATION

a	a constant
A	area
C	capacitance
d	thickness
D	diffusion coefficient
f	frequency
f_m	characteristic frequency
g	conductance of a non-linear element
G	conductance of a linear element
I	current
j	$\sqrt{-1}$
J_s	material flux
k	Boltzmann's constant
K	dielectric constant
l	electrode length
m	a parameter related to the width of a polarisation frequency spectrum
N	concentration
R	resistance
R_∞	resistance at high frequency
R_0	resistance at low frequency
R_s	'lumped' resistance
T	temperature
V	voltage
X	reactance
X_s	'lumped' reactance
Z	impedance
Z_p	polarisation impedance
α	$1-m$
β	$2\phi/\pi$
ϵ	permittivity
ϵ_∞	high-frequency permittivity
ϵ_0	low-frequency permittivity
ϵ^*	complex permittivity
ϵ'	in-phase permittivity
ϵ''	out-of-phase permittivity
ξ	energy gap between conduction and valence bands: activation energy
τ	relaxation time
τ_p	mean relaxation time
ϕ	phase angle
ω	angular velocity

6

ENVIRONMENTAL EFFECTS

Since skin is exposed to heat, cold, water, chemicals, radiation, friction and pressure, it must be durable and at the same time pliable enough to permit both gross and delicate motions of the body. Man has added greatly to the number of noxious, irritating and potentially dangerous agents in his environment; chemicals in the form of particulate matter and gases from industrial wastes now fill the air. The by-products of fertilisers, spray cans, and supersonic and upper-atmosphere flight may within the next few decades so affect our atmosphere that we notice changes in the spectral distribution of sunlight reaching the earth's surface. Added to this are the results of man's increased demands for energy, which include potentially harmful physical factors such as ionising radiation, microwaves, noise and increased thermal load. Even the enormous advantages to be gained through the introduction of lasers to world-wide communication systems will have to be balanced against the possible damage when impinging directly onto the human body or other living organisms.

CELLULAR DAMAGE BY ENVIRONMENTAL AGENTS

Cells of the skin exposed to environmental agents exhibit responses ranging from temporary disturbance in function, as increased or decreased activity, to transformation, malignant growth and death.

When carrying out experimental studies, better control of the effects can be obtained in cultured cells, but this limits interpretation and many of the specialised cells are excluded by lack of suitable culture technique. Despite this, the principal consequences of attack by environmental agents at the cellular level can be described.

When cells are *killed* by environmental agents the toxic effects will be either (*a*) threshold killing, (*b*) exponential killing, or (*c*) combined threshold–exponential effect. Threshold killing occurs when an agent which acts as a poison causes irreversible damage to a vital cellular function such as molecular synthesis or one of the membrane transport systems. The toxicity has unique characteristics: below a given concentration few effects will be detected, but then over a narrow dose range *all* cells will die. Exponential killing agents produce random lethal lesions within the target cell population or induce death randomly among cells of large populations in proportion to dose. They induce cell death over all dose ranges. The number killed is related to dose by a Poisson distribution. Chemical substances that cross-link DNA often give rise to an exponential response.

Threshold–exponential (TE) killing has characteristics of both of the above types of toxicity. At low dose levels, killing is minimal, but once the dose has reached a certain level death is exponential. Death rate of cells exposed to ultraviolet light (UV-C, 254 nm) is typical of this type of toxicity. There is one problem that persists, however, in relation to TE toxicity despite the volume of work reported: does death follow low dose levels of treatment? It is complicated by the need to distinguish between natural cell death rates and the response of cells to lethal doses. Few cells die immediately after receiving a fatal dose; some continue to function, even divide. The implication of this continued division of cells is important when considering the basal region of the epidermis where cells receiving the fatal dose may still produce terminally differentiating keratinocytes before death. Indeed, there may be a considerable time lapse between receipt of fatal dose and onset of detectable tissue degeneration.

Damage to cells may not of itself cause death, but result in mutation via changes in DNA production. The pattern of mutagenesis and mechanisms of DNA repair that influence the process are still not clear. However, there is strong evidence which links mutagenesis with cytotoxicity and suggests that a cell can tolerate only a limited mutational load before being inactivated.

Measurement of relationships between the cytotoxicity of an environmental agent and mutation has become a potent tool for predicting mutagenic hazard and the possible role of error-prone repair. DNA sequences susceptible to mutation (hot spots) are being identified in prokaryotic-DNA systems and at these spots the probability that specific agents induce various types of mutational events, such as specific transition or frame shift, is high.

Transformation of cells to malignant growth does show a correlation with cytotoxicity, but timing of dose in relation to cell reproductive cycle is important. That pre-malignant proliferation be required for the expression of the malignant state may be anticipated since cells already through the S-phase of cell division (prior to mitosis) are more likely to yield normal cells or daughter cells that die without completing further divisions. Again, one would expect a marked increase in transformation to occur at the point where exponential cytotoxicity begins in TE responses. This is well known for both ionising and non-ionising radiation.

Repair is now considered to be an important cellular response and reports on DNA repair in human fibroblasts suggest that at least three independent mechanisms can be utilised. Briefly, the *in situ* reversal of damage may be induced by (*a*) photoreactivation, (*b*) excision resynthesis, and (*c*) restitutional activities. There may be two types of excision resynthesis: (i) incision by an endonuclease of the DNA 'backbone' near the damage, followed by resynthesis of the section cut away, or (ii) cutting away of the sugar moiety and resynthesising. In general, it is assumed that the repair system invoked is determined by the structural nature of the damage to the DNA and that the damage itself somehow commands an increased local enzymatic activity. These concepts have many clinical implications, but more detailed and accurate knowledge from studies *in vitro* is required to provide a definitive basis for any clinical interpretation.

AGEING REACTIONS

Although not the primary object of study in most experimental work on skin, the reactions that affect or appear to affect the ageing processes are always of some concern. A relationship between changes in dermal structural composition and the development of malignancy in the epidermis has long been advocated (Gilman *et al.*, 1953; Mackie & McGovern, 1958). Alterations in appearance of skin with age occur over the whole body, but there are significant differences between regions exposed to solar radiation and those that are largely unexposed. The appearance of dry wrinkled skin with spotty pigmentation and telangiectasia (dilated blood vessels) was described by Unna in 1896 as 'sailor's' or 'farmer's' skin because the changes were seen predominantly on the exposed parts of the body of outdoor workers. The condition is not seen on covered skin and there is nearly always a clear demarcation.

In sections of skin from elderly patients, an increase in elastin tissue is sometimes called 'diffuse cutaneous elastic hyperplasia' or 'senile elastosis'. This term 'elastosis' is most unfortunate since it should mean a condition of being elastic, but it has been demonstrated over and over again that senile skin has fewer elastic fibres and is less resilient than young skin. Obviously, there is an increase in tissue in elderly people which takes up 'elastin stain', but this does not necessarily mean an increase in the skin's elasticity. Other factors relating to ageing processes will be discussed in each of the following sections.

CHEMICAL AGENTS

SURFACTANTS

Surveys have shown that 90% of women believe their primary skin problem is dry skin. They seek moisturising creams and lotions for relief with the understanding that 'moisturising' is an active word – to increase the water content of skin. However, there are many factors which affect the water content of the epidermis (see Chapter 1). Fatty occludants and humectants by preventing or inhibiting the evaporation of water from the epidermal surface provide benefits in the treatment of dry skin and so it would appear advantageous to combine them with water as an emulsion.

Emolliency, meaning soft and supple, is frequently taken to mean also pliable, lubricated and flexible. To a large extent the emollient state is due to fatty lubricants, yet it is not unusual to find anhydrous lipstick, eye cosmetics and lubricating night creams all claiming to have moisturising properties. Any improvement in water content is due entirely to the inhibition of water loss not to the addition of water. Petroleum products being occlusive and hydrophobic can be considered as useful 'moisturisers', but they are also greasy which makes them socially unacceptable. Indeed, there seems to be an inverse relationship between the effectiveness of such ingredients as occludants and their cosmetic feel.

When suitable petroleum products are made up as an emulsion, surface-active additives are needed to stabilise the water–oil interface and to retard loss of water by evaporation when applied to skin. These surface-active substances are normally anionic soaps or detergents; the cationic surfactants are seldom used.

Hand eczema was once attributed to soap used in industrial and domestic work, but in the last 30 years or so synthetic detergents have become the particular target. Their introduction has *not* led to a notable increase in housewife's eczema nor has the large-scale use of dishwashers and washing machines led to a conspicuous decline. Undoubtedly surfactants do influence the course of dermatitis, probably through cumulative aggravation, but many believe that this is mainly due to an enhancement of the effects of adverse climatic conditions, penetration of other irritants, of allergens such as perfume or metal ions, or of hormone disturbances within the systemic system.

Outside the laboratory, the causal relationship between surfactants and skin disorders has been difficult to prove. There are many contributory factors and laboratory tests remain the most satisfactory way to compare one product with another. Immersion tests have proved to be the best way to assess effects on living tissue. Smeenk (1968) found that a 1% solution of detergent frequently caused severe skin reactions when normal subjects were asked to immerse their hands and arms in the solution for 30 minutes each day. However, changes in temperature of the solution and changes in the general environment led to a wide range of results. Using a 0.1% detergent solution Smeenk determined by means of a scoring procedure the relative order of irritancy of six commonly used cosmetic or domestic products. The most irritant was coconut oil in the dimethyl amine oxide form, then sodium lauryl sulphate, secondary sodium alkyl sulphate, sodium alkyl benzene sulphonate, pure fatty acid soap flakes and iso-octyl phenol condensed with ethylene oxide. While soap flakes had the highest pH, the highest score was obtained from a detergent with a pH of 6, so defeating the then commonly held idea that pH of the detergent was the most important factor.

The detergent might be expected to cause the removal of fats, including those lipids bound into membranes, and to change the water-holding capacity of other substances. But solvents which remove free surface fats fail to have a lasting effect since the lipids are restored within a few hours. Temperature is probably of greater significance in everyday environmental effects on skin since lipids stiffen below 20 °C and flow or movement will be minimal. At elevated temperatures lipids show polymorphic changes and, in membranes, probably become more labile (Finean & Millington, 1955). After prolonged treatment resulting in the removal of lipids subsequent immersion in water will remove as much of the water-soluble substances from the keratin layer in just 2 minutes as a 2-hour immersion in tap water not preceded by defatting (Idson, 1973). It seems, therefore, that there is a complex relationship between extracellular fat, cell lipids, water content and temperature. The temperature factor is further exemplified by the failure to produce housewife's eczema in summer unless the skin is immersed in ice-slush and acetone.

Surfactants probably do not penetrate skin but merely affect the stratum corneum. Blank & Gould (1961) failed to demonstrate penetration of the epidermis by synthetic anionic surfactants provided the natural epidermal barrier remained intact. Blank, Gould & Theobald (1964) later showed that the same was true of cationic surfactants. Penetration, however, was rapid if the skin had been in contact with alkalis or lipid solvents, or had been stripped. In studies *in vitro* water permeability is dramatically increased in the presence of surfactants such as sodium laurate and sodium lauryl sulphate (Scheuplein & Ross, 1970). The effects of surfactants continue to increase with time and do not attain a steady state.

Using 5% surfactant (pH 10.2) the permeability constant for water ($k \times 10^2$ cm/h) reaches a maximum value of 88 after the skin has been in contact with the detergent for 4 hours, an increase over the normal rate of some 450 times. With 1% surfactant, the permeability constant reaches a maximum of 61 after 6 hours in contact, but decreases in value to 3.2 some 55 hours after the treatment has been stopped. The presence of a detergent increases the permeability of skin for quite a long time and, by comparison, treatment by soaking in buffer salts, sodium sulphate or dilute acetic acid has little or no effect.

From these and many similar experiments it would seem that the effect of detergents on skin is largely related to changes in permeability to water and not to diffusion of the detergent into the dermal regions. This explains why many workers who have their hands in water for long periods of time in the presence of dilute soap or detergent solutions do not develop skin disorders although the skin may be extraordinarily hydrated at the end of each working session.

METAL IONS

Either through direct contact or from airborne particles the skin comes into contact with a variety of metal ions or salts many of which induce toxic reactions. Copper, arsenic, mercury, cadmium, nickel and aluminium have all been identified in toxic quantities in skin and hair. In a wide variety of industrial processes salts of metals in solution have been directly associated with dermal disorders among workers. Certain cosmetics contain metal salts in quantities sufficient to cause skin reactions. Absorption of a metal ion increases with concentration until a maximum value characteristic for each salt has been attained. Thereafter, absorption generally decreases (Skog & Wahlberg, 1964). This property is readily explained by the binding of metal to protein.

Tests for sensitivity to metals either in the form of salts or as pure metal can be made by patch tests where a small area of skin is subjected to contact with the metal or where small quantities of soluble salts at different concentrations are injected into or just below the epidermis. Some metal ions penetrate the skin rapidly and then give rise to systemic disorders and in these cases cutaneous adverse effects are regarded as manifestations of hyper-sensitivity (Kennedy *et al.*, 1977).

Skin disorders arising from metal toxicity include erythema, urticaria, exfoliative dermatitis, erythema multiforme and acrodynia (pink disease). The red areas of tattoos are sometimes accompanied by dermatitis caused by cinnabar (HgS). When a metal lies in skin in particulate form it is normally taken up by phagocytes (macrophages), but some metals can become associated with other cells or extracellular materials. Where metal compounds are being exploited as astringents, antiperspirants, antacids or antibacterial agents, severe reactions can arise with hyperplasia and focal ulceration. Such extreme reactions are common in sensitive people. In general, all metals should be treated as potentially toxic to the skin whatever the form of presentation.

GASES

The permeability of skin to gases is very high, but despite the potential for toxicity few reports can be found. It is important to distinguish between true gases, such as the inert gases, oxygen, nitrogen, hydrogen, etc., and the vapour-phase concentrations of volatile liquids and solids. Aerosols penetrate the skin as liquids or even as aqueous solutions.

Oxygen is absorbed from the air and carbon dioxide is excreted from skin at near equal rates. The epidermis acts as though it is oxygen deficient, absorbing at very low partial pressures, and the outward flow of oxygen is negligible, even in an atmosphere of pure nitrogen. Diffusion through the stratum corneum seems to be the rate-limiting process as evidenced by the vastly increased rate of gas exchange when the corneum is removed.

The rate of transfer of carbon dioxide is important particularly in view of the increased use of lasers for the removal of tattoos, where the laser energy transforms the carbon moiety of the dyes into carbon dioxide. Carbon dioxide excretion is very temperature dependent, but the changes with temperature cannot be attributed solely to diffusion processes since metabolic activity involving the release of carbon dioxide is also coupled to temperature changes. Nevertheless, the diffusion constant for carbon dioxide is about 200 times greater than for water but about 3 times slower than for oxygen (Shaw, Messer & Weiss, 1929).

Of the inert gases, helium diffuses very much faster than nitrogen or argon. Activation energies are large, and this is consistent with the absence of pores or air-filled interstices through which gases could diffuse. Inert gases are appreciably soluble in oil and their permeability through the stratum corneum is accordingly larger than might be expected from simple consideration of partition coefficients with water. The popular myth that gases pass readily through skin is only partly true. It would seem that rates of flow would be about 100 times greater than for liquids. The total quantities transported, however, are no greater for gases than for pure liquids and dilute aqueous solutions, although the time for transport is measured in seconds rather than minutes. The diffusion rate for these gases lies between 2 and 21×10^{-8} cm²/s, whereas that for water is about 3.1×10^{-10} cm²/s (Scheuplein & Blank, 1971).

Vapours of alcohols and hydrocarbons diffuse through skin dissolved as aqueous solutions. It follows that those with low solubility have much lower diffusion constants than water. Thus, the diffusion constant for alcohols and alkanes decreases with increasing molecular weight and the presence of hydroxyl groups is also associated with decreasing rates. C_{10} and C_{14} alkanes may have measurable permeability constants greater than water, but otherwise the measured constants are all equal to or less than that of water.

Because of the higher diffusion rates for permanent gases compensating for higher surface concentrations of vapours, it is difficult to demonstrate higher transport rates between the two groups. In both cases the stratum corneum appears to be the primary controlling factor. In very thin skin transport rates are always relatively slow.

HORMONES

Many preparations applied to skin contain hormones. These are usually of the corticosteroid type or the sex hormones. Corticosteroids are widely used for a variety of dermatoses because of their anti-inflammatory activity, but when used repeatedly give rise to unwanted side effects, the commonest of which is atrophy. Atrophic changes can manifest themselves in a variety of ways such as striae, wasting of sub-cutaneous fat and muscle, bruising, skin fragility or as 'simple' thinning. Some changes are not reversible and topical application of steroids may aggravate existing clinical conditions such as rosacea, mask fungal infections and encourage bacterial growth. Patients with chronic dermatoses sometimes discover that withdrawal of treatment leads to a major relapse; for example, withdrawal from patients with psoriasis has been known to precipitate the pustular form of the disorder.

Prolonged applications of corticosteroids have been known to lead to contact allergies and, in children, a case of death from adrenal failure following withdrawal of treatment was reported by Sneddon (1976). Nevertheless, the use of corticosteroids in the treatment of skin disorders continues because nothing better has yet been found.

The choice of corticosteroid in the treatment of skin disorders is complicated by the lack of information about the dose of drug absorbed and the frequency and quantity used when applied by the patient during treatment. Synthetically produced analogues may enhance certain effects or suppress them and much effort has to be made when investigating the properties of modified corticosteroids to avoid some of the more troublesome side effects. Many experimental models have been proposed in attempts to predict the efficiency and potential of these preparations, but the differences between animal and human skin make some of the systems difficult to interpret.

Several bio-assays have been developed to predict skin atrophy, which is the most important side effect. Atrophy itself is commonly assumed to result from an antimitotic or antisynthetic effect of the corticosteroid. Thinning of the epidermis may result from changes in the size of cells rather than the number of cells present (DelForno, Holt & Marks, 1978). Reduction in cell size may precede the changes in mitotic activity, which have been well established (Fisher & Maibach, 1971). This being so, corticosteroids may affect macromolecular synthesis within the keratinocytes, a suggestion supported by alterations in ribosomes and endoplasmic reticulum in steroid-induced atrophic skin (Stefanovic, 1976).

Direct measurement of skin atrophy has only limited value (Dykes & Marks, 1979). The major errors, whether by micrometer or Harpenden skin-fold calipers, derive from the uncontrollable changes in thickness under the pressure applied to the skin fold. Since the sub-cutaneous fat layer is also included in this type of measurement compensation errors for the thickness of fat add further to the difficulties in obtaining reliable information. Techniques used to overcome some of these restrictions include gravimetric, ultrasound, soft X-ray, histiometric and certain indirect methods including measurement of electrical resistivity. All are constrained in practice by the inability to control or measure changes in water content which affects not only the measurements themselves but also the rate of absorption of drug into the tissue.

Radiographic measurements of skin thickness after treatment with various corticosteroids have indicated that significant changes in skin thickness can be induced over a period of 4 months. However, not all steroids induce changes. Hydrocortisone-17-butyrate, flurandrenalone and clobetasone butyrate do not induce significant changes, but fluocinolone acetonide does in all cases (James, Black & Sparkes, 1977). This was confirmed later by Dykes, Marks & Blakemore (1978) who carried out a double-blind study and failed to show that hydrocortisone-17-butyrate and betamethasone-17-valerate, applied in a cream base, could produce significant atrophy over a 6-week period. Of the commercially available corticosteroids only clobetasone butyrate failed to cause epidermal thinning (Wrench, 1980).

The most dramatic changes in skin occur in the first week of application of steroid and this has led to the suggestion that it is the glycosaminoglycan metabolism that is being affected. If so, then water content is also affected. It is probable that collagen is relatively unaffected.

Hydrocortisone has been known for some time to be involved with the adrenaline–adenylate cyclase system in skin. The role of adrenaline and cyclic AMP particularly in association with chalones has been described in Chapter 3. In experiments on pig skin the accumulation of cyclic AMP following treatment with various combinations of hydrocortisone and adrenaline has been measured by Hzuka *et al.* (1980). Using radio-immunoassay for cyclic AMP phosphodiesterase they demonstrated that in the short term neither hydrocortisone nor adrenaline, separately or in combination, had any effect on the accumulation of cyclic AMP, but the accumulation was marked if incubation in hydrocortisone was pursued for 6 hours or more. In particular they noted that if skin was incubated in hydrocortisone for 48 hours and then with 50 M adrenaline for 5 minutes, the amount of cyclic AMP was concentration dependent, the maximum effect being observed with a concentration of 10 M hydrocortisone. The amount of cyclic AMP measured at the peak concentrations was over 40 pmol/mg protein whereas without hydrocortisone the level was less than 3 pmol/mg protein. It seems, therefore, that hydrocortisone acts by protecting the adrenaline–adenylate cyclase system and in doing so has a direct effect on the mitotic rates of cells in the skin.

In 1946, M. Goldzieher reported that topical application of oestrogens to skin of elderly women with advanced senile atrophy of the skin produced regenerative changes. The number of cell layers was greater and the 'elastin' fibrils were more numerous (see also Chapter 4, p. 94). J. Goldzieher (1949) later found that oestrogen did not induce regeneration in men when applied in a cream base, but when oestrogen or androgen was sprayed onto the skin in sufficient quantities, regeneration of skin occurred regardless of sex. Appraisal of the effects of the steroids was made from histological sections of biopsy samples. As a result of these studies and similar reports (Eller & Eller, 1949) steroid hormones have been widely used in cosmetics to improve skin texture.

Human hair, particularly axillary and pubic hair, is androgen dependent. The associated sebaceous glands are also stimulated by androgens and inhibited by oestrogen and anti-androgenic steroids (Ebling, 1974). Reduction of the androgens by 5α-reductase converts them to the active form, and the measurement of this enzyme is an important indicator of sites of androgen activity. The highest activity of 5α-reductase is found in the axillary sweat glands of both men and women. Under standardised conditions values as high as 400 pmol/mg dry wt per hour have been found (Takayasu *et al.*, 1980).

At the same site, values in sebaceous glands varied from 85 to 261 pmol/mg dry wt per hour. Hair follicles themselves have much less activity and the quantity in the epidermis is negligible.

Progesterone inhibits 5α-reductase activity and when applied topically in women inhibits sebum secretion (Simpson *et al.*, 1979). It appears that, at least in women, dihydrotestosterone is an important sebotrophic factor. After topical application of progesterone to the skin, formation of 5α-dihydrotestosterone is inhibited (Vermorken, Goos & Roelops, 1980). If applied to one side of the scalp inhibition occurs also on the contralateral side indicating either a systemic effect or a local diffusion of the steroid. Skin is, therefore, a major site for androgen metabolism comparable to the prostate gland and it is the 5α-dihydrotestosterone formed in the skin from testosterone that is the active form (Price, 1975). The clinical implications are important particularly in the case of acne where increased local formation of 5α-dihydrotestosterone is well documented. Treatment of these disorders, however, is not simple because of the side effects following administration of steroid hormones.

THERMAL RESPONSES

Application of heat to skin induces first discomfort then pain and finally burn injury with death of the adjacent tissue. Clinically there are two levels of burn: those that involve partial thickness of skin and those in which the entire depth of skin including the underlying tissue has been destroyed. Superficial burns can heal by regrowth from surviving epithelial elements. Deeper burns will heal through ingrowth of new epithelium from the edges of the wound. The third vascular plexus will re-form quite quickly. Damage to the full thickness of skin leads to unsatisfactory healing and is often treated by skin grafting even when the damaged area is only a few centimetres in diameter. The combination of time and temperature to produce these different types of injury has been explored by Moritz & Henriques (1947) and standard curves showing minimum time and temperatures to induce such burns have been published by Bull (1963). Typical relationships between time and temperature to cause discomfort and thermal injury are shown in Fig. 45.

Discomfort is a highly subjective feeling which may be reported as pain, throb, smart, sting or ache, but is generally sensed by the hand at temperatures of about 43 °C. The time an object above this temperature can be held is inversely related to the temperature. Immersion in liquids usually lowers the level at which discomfort and then intense pain are felt, but individual variations are high and it is not uncommon to find workers who can immerse their hands in water at 60 °C without apparent distress or superficial burning. The use of gloves in both industrial and domestic tasks leads to the adoption of higher working temperatures and hence increases the risk of accidental burns.

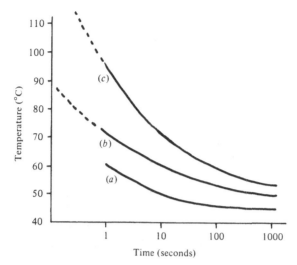

Fig. 45. The relationship between time and temperature for (*a*) discomfort, (*b*) superficial burns (partial skin thickness) and (*c*) full-thickness burns.

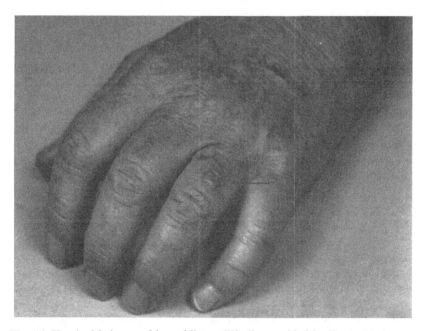

Fig. 46. Hand with 3-second burn blister. (Kindly provided by Dr A. M. Stoll.)

TABLE 9. *Constants used in calculation of maximum permissible temperature of a material to onset of pain*

Contact time (seconds)	m_1	k	Material	$T_p = 1/\sqrt{k\rho c}$
1	1.094	0.490	Aluminium	2.0
2	0.963	0.368	Steel	4.5
3	0.894	0.312	Heravit	22.0
4	0.848	0.276	Glass	32.0
5	0.814	0.252	Teflon	56.5
			Masonite	78.0
			Polypropylene	93.0

After Stoll, Chianta & Piergallini (1979). For details see text.

The time between initial contact and the sensation of pain or the later initiation of blistering (Fig. 46) depends upon the thermal properties of the object and the thickness of the epidermis, not merely on the temperature. The thermal properties of any material can be defined as the square root of the reciprocal of the thermal inertia, thus:

$$T_p \text{ (thermal property)} = 1/\sqrt{k\rho c}$$

where k is thermal conductivity, ρ the density of the material and c its specific heat. Values for T_p are given in Table 9, for a selection of materials.

If the temperature of each material at the onset of pain in a fixed time (seconds) is plotted against T_p the relationship is linear (Stoll, Chianta & Piergallini, 1979). This maximum temperature of a material to give pain in pre-determined time is designated by 'TmPT'. Stoll and her colleagues showed that

$$\text{TmPT} = m_0(T_p + 31.5) + 41$$

where m_0 is given by the relationship

$$\log(m_0) = m_1(s) + \log k$$

s being the thickness of epidermis, and m_1 and k experimentally determined constants given in Table 9.

It is possible to calculate the value of TmPT for any material for a given exposure time up to 5 seconds provided the thickness of the epidermis is known.

Example: To find the maximum temperature of glass that will induce threshold pain at an exposure of 4 seconds with epidermal thickness 0.43 mm.

From Table 9 we take values for m_1, k and T_p. Then from the above equations:

$$\log(m_0) = 0.848\,(0.43) + \log(0.276)$$
$$m_0 = 0.64$$
$$\text{TmPT} = 0.64\,(32 + 31.5) + 41$$
$$= 81.6\,°C$$

This computational system is useful for those who select materials for use in construction and manufacturing.

Heat-damaged skin has proved extremely difficult to assess. Little coagulation of tissue protein takes place at temperatures below 60 °C and at lower temperatures tissue simply cools by re-radiation and by conduction of heat by the vascular system. The efficacy of the vascular system in conducting heat is well exemplified through experience when branding animals. While the surface of skin may be raised to 100 °C or more by the iron, at a depth of 1.5–2.0 mm below the surface the temperature may not rise above 30 °C (Cruikshank & Hershey, 1960). A temperature of more than 50 °C for at least 60 seconds is needed to damage isolated skin. However, enzymes can be damaged at temperatures well below that needed to damage collagen. Changes in cellular respiratory activity occur at temperatures above 40 °C in experiments both *in vitro* and *in vivo*, with more than 75% reduction in activity at 47 °C. Clearly, significant changes can occur long before clinically recognised burns are reported.

The situation is further complicated by the fact that skin cells adjacent to heat-coagulated tissue may not become necrotic until several days after injury. To understand the problem better, attempts have been made to assess damaged areas more accurately by defining relationships between pain threshold and thermal radiation (Stoll & Greene, 1959). To do this an estimation of tissue damage must be made. Moritz & Henriques (1947) using constant temperature levels derived an equation relating the exposure time, rate of protein denaturation and temperature:

$$\Omega = P \cdot e^{E/RT} \cdot t$$

where Ω is tissue damage, P is integration constant, E is energy of inactivation, R is the gas constant, T is tissue temperature and t is exposure time.

Tissue damage was defined as the integral of damage rate $(d\Omega/dt)$ for the exposure time noted and was given a value of 1.0 when blistering of skin occurred. Pain threshold for any given radiation intensity appears at about one-third of the time of blistering. At 400 mcal/cm² per second this was about 2 seconds with blistering at 5.6 seconds. At 100 mcal/cm² per second pain threshold occurred after 12 seconds and full blistering after 37 seconds.

The relationship between radiation intensity and pain or blister threshold is not linear (Fig. 47). The damage rate versus temperature curves show a logarithmic relationship, but with a change of slope at 50 °C when plotted

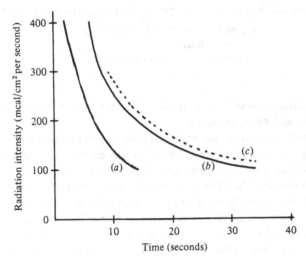

Fig. 47. The relationship between time and radiation intensity to (*a*) pain threshold, (*b*) blister threshold and (*c*) deep blistering. (After Stoll & Greene, 1959.)

to log scale. Conductive heating and pain intensity increase with temperature uniformly as the logarithm of the damage rate.

Sensing changes in skin temperature is important. A trained observer can resolve remarkably small incremental changes in skin temperature if the changes occur rapidly. The subject reports warming or cooling of his skin and qualifies this in terms of its intensity and duration. He also reports clear-cut changes in the character of the sensations until pain is perceived provided the applied temperature rises above 45 °C or below 15 °C. Man also has a wide range of pain sensation. He can distinguish 21 levels of pain as the tissue temperature rises from 45 °C to 56 °C and it is possible to correlate the temperature of the epidermal basal layer with these pain levels (Hardy, Wolff & Godell, 1952). Subjects become much less sensitive if the changes in temperature are slow.

In order to detect a change in temperature there must be a minimum time to allow discrimination (Darian-Smith & Johnson, 1977). This time period is made up of two components: the peripheral neural processing (transmission time) and central neural processing including comparison of incoming signal with the memory of the previous signal level. In monkeys three distinct nerve groups have been found in glabrous skin associated with temperature discrimination: (1) 'cold' fibres, which are uniquely excited by cooling the skin, are suppressed by warming and unresponsive to mechanical stimulation; (2) 'warm' fibres, excited by warming the skin, suppressed by cooling and unresponsive to mechanical stimulus; (3) many of the slowly adapting type A mechanoreceptor fibres, which respond to local cooling but are less sensitive to rapid cooling than the 'cold' fibres. It is generally assumed that a similar system operates in man.

No single fibre carries sufficient information to account for the human subject's ability to resolve incremental changes in stimulus intensity (Darian-Smith & Johnson, 1977). This has been taken to imply that some integrative process within the central nervous system combines information signalled by individual fibres to provide more precise stimulus information. It has been deduced that 20 or more fibres are needed to provide sufficient signal to match human intensity resolution. Clearly, the advantages of central processing enable the brain to maximise the information from the more sensitive or responsive fibres and downgrade the effect of noisy fibres.

While skin contains 'cold' fibres to signal decreasing temperature, loss of tactile sensitivity at low temperatures is a well-recognised phenomenon. Slower growth rates for hair and nails have been assumed since McLean (1919) and later Le Gros Clark & Buxton (1938) reported measurements on growth rate changes in the Antarctic and British winters respectively. In more recent reports, measurements taken on Antarctic expeditions have indicated slightly less reduced growth rates than those reported by McLean, but this is probably due to better living conditions (Donovan, 1977). Despite the fact that in these and many other studies the actual finger temperatures were not measured, it is still assumed that lack of sensitivity is due to cold.

One of the difficulties in assessing any changes due to cold is the level of adaptation of the subject and this is further complicated by the difficulty in finding subjects who have spent sufficient time in sub-freezing conditions to achieve maximum change. Structural changes in skin have been measured on a team who spent a year with the Australian National Antarctic Research Expedition (Bodey, 1978). The mean values of skin thickness were taken from biopsy specimens fixed in formol saline. The results showed a highly significant increase in epidermal thickness ($P < 0.001$). A much more significant result, however, was obtained by taking the mean number of epidermal cells per unit area. On both the dorsum of the hand and on the abdomen there was an increase in number of cells, although the increase on the abdomen, a well-protected area of the body maintained at a much higher temperature, was only just significant at the $P < 0.05$ level. The cell density on the dorsum of the hand, however, increased by 37.5.%. Bodey also demonstrated an increase in the number of layers of keratinised cells in skin, but this could have been due to changes in work conditions as well as changes in cell turnover rate. This result appears to be contrary to the expected changes in skin from cold, which were assumed to reduce not only sensitivity but also cell activity. Presumably the hands were kept warm enough to prevent supercooling and frostbite.

Skin seldom freezes without some degree of supercooling. On immersion in brine at $-0.6\,°C$ true freezing was achieved only if the fingers were surrounded with several layers of plaster to control the rate of heat loss

(Keatinge & Cannon, 1960). In the presence of wind, supercooling always occurs and there is a measurable rise in skin temperature as ice crystals form. In a series of experiments in which the hands of volunteers were exposed to air at temperatures below freezing and different wind speeds, it was found that more than half the volunteers did not register frostnips. This was due largely to cold-induced vasodilation (Wilson, Goldman & Molnar, 1976). Of the remainder the highest air temperature at which freezing occurred was calculated to be $-15\,°C$ compared with $-13\,°C$ for unprotected hands immersed in brine.

The effect of cold injury to skin is not unlike that caused by heat burns. Regeneration depends upon the rate of mitotic activity and the rate of synthesis within cells. It is likely that many cells retain active protein which on warming can stimulate the tissue, whereas in heat burns the tissue is destroyed. Delay in healing is related to the amount of necrotic tissue that has to be resorbed. Hell & Lawrence (1979) have found a delay of up to 3 days in the number of cells entering DNA synthesis after heat burn. The normal maximal response after physical injury is seen after about 24 hours. Since frozen cells have to be fully digested and resorbed when they die, it is thought that cold burns heal more slowly than comparable heat burns. However, the available evidence neither supports nor contradicts this hypothesis.

SOLAR RADIATION

The sun's radiant energy sustains all life on earth. It maintains the earth's temperature, is essential to photosynthetic activity in plants and is involved in a variety of chemical reactions at or near the body surface in all animals including man. Our skin, eyes, blood vessels and certain endocrine glands all respond to radiation and biological rhythms are dependent upon the movement of the earth relative to the sun or cycles of sunlight. Yet excessive solar radiation is harmful.

The effects of radiation upon skin depend on the amount absorbed by the cells. Absorption is a relatively precise phenomenon. Specific molecular structures absorb radiation of a pertinent wavelength. In general, absorption of infra-red wavelengths increases the energy levels or temperature of the tissues. Because it is absorbed more readily by the denser connective tissues the heating effects of infra-red are used in physiotherapy to relieve pain, to increase fluid flow and influence muscle function. Light radiation (wavelengths 400–700 nm) is largely sensed through the retina. In practice most sources of visible light also emit infra-red and/or ultraviolet light. It is from these longer and shorter wavelengths that deleterious effects normally arise.

Ultraviolet radiation is subdivided into a number of bands related to their phenomenological effects. These subdivisions are somewhat arbitrary and vary according to the discipline involved. The three principal divisions, A, B and C, cover the wavelength range 200–400 nm. The range 200–290 nm is

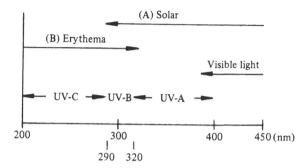

Fig. 48. The relationship between the three ultraviolet light divisions and associated phenomena. A, solar spectrum at the earth's surface; B, skin erythema (delayed) action spectrum.

usually known as UV-C, 290–320 nm is called UV-B and 320–400 nm is called UV-A (Fig. 48). Wavelengths shorter than 290 nm do not reach the earth's surface, being absorbed by the ozone layer surrounding the atmosphere.

Radiation in the UV-C band rapidly causes erythema of skin and can also cause photokeratitis (inflammation of the cornea). It is also effective in killing single-celled organisms and can be used as a germicidal radiation for sterilisation. UV-B reaches the earth's surface in relatively small quantities, but it is very effective in causing redness (sunburning or erythema) of skin. It is also thought to cause skin cancer as well as being responsible for the age-related changes in exposed skin.

UV-A has, until recently, been thought to be quite innocuous. It stimulates the reproduction of melanin granules, but can, in large doses, cause erythema. Certain types of cataract are thought to be caused by UV-A in large doses. Most domestic glass and many plastics screen out the incident UV-B without altering the amounts of UV-A significantly and so indoor workers may still receive large doses of UV-A. It is also stated that variations in incident solar UV-B exposure are responsible for fluctuations in vitamin D levels; but since the rate of synthesis in skin is small compared with the quantities ingested, it is probable that the effects are only crucial when dietary deficiencies obtain.

PENETRATION AND ABSORPTION BY SKIN

When radiation strikes the surface of skin some is reflected directly by the outer layers of the stratum corneum. This reflection will increase as the angle of incidence decreases and with the amount of fluid on the surface. Internal reflection from the various layers of the epidermis will further reduce the transmission into the dermis. Some of this scattered radiation will be absorbed, some will pass back out of skin and the remainder will pass into the dermis albeit at different angles to that of the incident beam. The

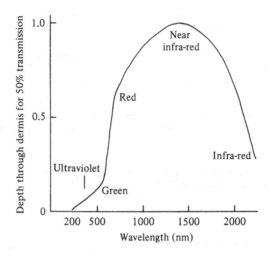

Fig. 49. Variation in skin penetration with wavelength plotted as the 50% transmission depth in skin.

quantities 'lost' at each level will be dependent also on wavelength. Very little ultraviolet radiation penetrates the dermis although visible light, particularly in the red and near infra-red region, is transmitted quite well (Fig. 49).

The absorption of radiation within skin can be attributed in part to specific substances: oxyhaemoglobin has strong absorption at about 420 nm, haemoglobin at about 450 nm and β-carotene at 450–520 nm. Much of the UV-A radiation, therefore, can be absorbed by components associated with the vascular system. Particularly at risk are the capillary vessels in the papillary plexus. Roughly 30–50% of the incident UV-A reaches the dermis, but significant amounts may pass on through to be absorbed by the fat layers (Everett *et al.*, 1966).

A single large dose (86–173 J/cm²) of UV-A from a xenon lamp is sufficient to cause pronounced dermal damage (Kumakiri, Hashimoto & Willis, 1977). In the papillary region the small dermal vessels are opened and extravasation of blood cells and platelet aggregation occur. The pericytes of the blood vessel walls are damaged and pyknotic nuclei can sometimes be found. Fibrin deposition between the collagen bundles may be found in the reticular dermis. The minimum erythemic dose (MED) to give a response 24 hours after irradiation has become a standard value to quote. For UV-A the MED is generally accepted as being 42.8 J/cm² although Kaidbey & Kligman (1975) suggest a value of 86 J/cm² and Parrish *et al.* (1974) suggest 20 J/cm². With less than two MED of UV-A radiation no damage to the epidermis occurs.

Similar levels of UV-B or mixed UV-A + UV-B radiation always give rise to epidermal damage as expected from the absorption data. By electron microscopy the damage is seen as dyskeratotic cells and other cytoplasmic changes.

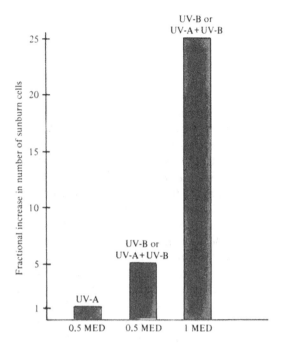

Fig. 50. Functional increase in the number of sunburn cells per unit area of epidermis after irradiation.

Large variations in MED values undoubtedly arise because of differences in experimental procedure. Delayed reactions are open to great subject variability and to overcome this problem Kaidbey & Kligman (1978) measured the acute effects of UV-A. Using healthy Caucasian students aged 19 to 24 years, they found skin redness appeared immediately after exposure to 50 J/cm² and persisted for 24 hours. Threshold doses of 13 J/cm² gave redness which faded within a few minutes. With doses of 18 J/cm² immediate pigmentation of the skin persisted until delayed pigmentation (true melanogenesis) took over without noticeable fading. Melanogenic efficiency is wavelength dependent and UV-B is more efficient in stimulating delayed melanogenesis. The minimum tanning dose (MTD) for UV-B is about 50% larger than the MED whereas with UV-A the MTD is about 400 times greater than that needed with UV-B.

One of the characteristic histological effects of UV-B is the formation of the so-called sunburn cells in the epidermis. These are assumed to be damaged keratinocytes. They have a pyknotic nucleus and eosinophilic cytoplasm. By counting the frequency of these cells it is possible to obtain a measure of acute epidermal damage. Very few sunburn cells are seen in normal unirradiated skin. After irradiation with UV-B the number of sunburn cells increases dramatically with exposure (Fig. 50). By contrast the increase after UV-A irradiation is not significant.

SOLAR DAMAGE

In addition to erythema of the skin and the production of sunburn cells there are other consequences resulting from overexposure to solar radiation. The relationship of sunlight to skin cancer has been known for about a century (Orth, 1887). Thickening of the stratum corneum has been observed as one of the features of the epidermal response to ultraviolet radiation. A thick parakeratotic zone forms beneath the normal stratum corneum within 72 hours of irradiation. After about a week post-irradiation a proximal corneal layer is formed beneath the parakeratotic layer of comparable thickness to the distal corneal zone (Thomas *et al.*, 1964).

Changes in cutaneous blood flow can also be measured. Although 90% of the incident radiation of wavelength 253 nm is absorbed by the epidermis, delayed erythema does arise and has been described by Levan *et al.* (1964) as biphasic. There is a decrease in blood flow immediately after irradiation with a return to normal flow shortly afterwards. Ramsay & Challoner (1976) measured by three methods the blood flow following irradiation at 250 and 300 nm. Using skin thermometry, photoelectric plethysmography and thermal conductance they were able to demonstrate reduced flow through the papillary plexus vessels which, they claimed, was probably due to stasis. However, there is a general increase in blood flow through skin after irradiation as demonstrated by the elevation of skin temperature for up to 96 hours.

Of great concern has been the possible changes in DNA within cells affected by ultraviolet light. The mutability of cells differs according to type but the phenomenon does have clinical consequences. Studies *in vitro* have shown that fibroblasts from xeroderma pigmentosum patients have a higher mutability than normal fibroblasts when irradiated with ultraviolet light (Maher & McCormick, 1976; Maher *et al.*, 1976). It appears that the DNA repair processes are affected in the disease so that premature solar damage occurs in all xeroderma pigmentosum patients (Robbins & Moshell, 1979). The corollary of this is that DNA repair processes protect human skin cells from premature solar damage.

It has been known for some time that furocoumarins (psoralens) when administered orally or topically increase the sensitivity of skin to ultraviolet radiation. The major interaction of the psoralens seems to be with the DNA and RNA of epidermal cells (Pathak & Kramer, 1969). The formation of the C_4-cycloaddition product 4,5′,8-trimethylpsoralen with the DNA or RNA appears to be responsible for the effects.

With PUVA treatment, that is treatment with psoralens prior to UV-A exposure, at least two types of DNA lesion are produced: photo-binding of psoralen to the DNA or dimerisation of pyrimidines. These DNA lesions are subject to excision repair (Hanawalt *et al.*, 1979). Using autoradiographic techniques the *de novo* synthesis of DNA segments needed for the repair can be detected. However, it is likely that it is not, in fact, the UV-A that causes the DNA damage. Normal commercial UV-A units emit a small quantity of UV-B and it is this that causes damage at low overall levels of exposure. As dose levels are increased repair activity also increases linearly, indicating that repair is normally very efficient in human skin (Bishop, 1979).

PUVA therapy has been used with success in the treatment of psoriasis (Melski *et al.*, 1977). Filtered radiation in the range 320–400 nm (UV-A) was used 2 hours after administering 0.6 mg/kg 8-methoxypsoralen by oral ingestion. The treatment was given to inhibit the overall DNA replication and cell multiplication in the hyperproliferative plaques. The success of the therapy indicates that there is a synergism between UV-B, UV-A and psoralen although the precise mechanism has yet to be clarified.

Recent studies on the histological changes in skin following irradiation with either UV-C, UV-B or UV-A only served to confirm that UV-A and psoralen effects are related to the greater penetration of longer wavelengths into skin. The PUVA effects are longer lasting (Rosario *et al.*, 1979).

One of the possible consequences of DNA damage is the production of carcinogenic tissue. Both crude coal tar and ultraviolet radiation are used in the treatment of psoriasis and both are known to increase the risk of cutaneous cancers in animals. An estimated relative rate of cutaneous carcinoma for patients with high exposure to coal tar or ultraviolet radiation or both has been found to be 2.4 times greater than for patients with low exposure histories (Stern, Zeirler & Parrish, 1980). Using a matched series of 126 patients the association was 4.7 times greater. In any event, given early detection and treatment of cancer of the skin, this risk is not taken as a contra-indication particularly when compared with the long-term physical and psychological effects of psoriasis.

In addition to the photosensitisation caused by the psoralens other chemical substances and drugs when applied to the skin may cause photo-dermatitis. A number of cases were reported by Wilkinson (1961) when tetrachlorsalicylanilides were added to soap. Since then a number of substances added to soaps have caused similar problems. Patients suffering such disorders fall into two categories: (*a*) those who recover as soon as the active substance is withdrawn, and (*b*) those who continue to be photosensitive, the 'persistent light reactors'. The reason for persistent photosensitivity is unknown. It is frequently a general condition but local sensitivity can occur. The apparently normal skin of 'persistent light reactors' continues to show abnormal reactivity to UV-B and UV-A.

After investigating four patients who became 'persistent light reactors' following contact dermatitis from Fentichlor, Ramsay (1979) concluded that the sensitivity was not spurious. Three of the patients had come into contact through hair cream containing Fentichlor and the fourth from industrial exposure. All four patients showed increased sensitivity to both UV-B and UV-A. This could have been due to the self-absorption properties of

Fentichlor, which at pH 4 has an absorption peak at 320 nm, at pH 7 a peak of 290 nm and at pH 9 a peak at 320 nm. On the other hand, there is no evidence that drugs affecting sensitivity remain in the tissue in any appreciable quantity once the source has been removed.

SUNSCREENING

Screening the skin from the harmful effects of solar radiation appears to be a good idea, particularly when the incidence of basal and squamous cell carcinoma and malignant melanoma is doubling every 10 years or so. More than 50% of all human cancers are those affecting the skin, although this statistic may be a little misleading since most skin cancer is easily cured. A large number of Caucasian people simply do not tan, they are all cancer-prone, they burn readily and heal slowly or badly. These are the type 1 people. Type 3 and 4 people have skin that tans rapidly and rarely burns. Yet only in recent years has it been realised that those who tan well can still show delayed effects of radiation damage. While we can feel infra-red radiation through temperature sensors in the skin, ultraviolet radiation is below the sensory perception level. Overexposure to UV-B can lead to carcinogenesis in all types of skin.

By altering optical and photochemical pathways many substances, whether applied topically, introduced systemically or produced endogenously, can give protection against the damaging effects of ultraviolet radiation. The arbitrary choice of a light-screening agent without first of all correlating its absorption spectrum with the action spectrum responsible for the skin reaction has led to the incorrect assumption that light-screening agents are of little use in routine treatment of photodermatoses.

Most of the agents employed as solar sunscreens are topically applied chemicals. *Para*-aminobenzoic acid or its ester amyldimethylaminobenzoate have absorption spectra mainly confined to the shorter wavelengths and therefore give protection against UV-B. Benzophenones and benztriazoles absorb over a wide spectral range including some portions of the UV-C and UV-A bands. Some diphenylketones absorb appreciably in the wavelengths longer than 320 nm, and the cinnamates and salicylates absorb in the UV-B region.

Since *para*-aminobenzoic acid is soluble in alcohol it can be applied in high concentrations and when combined with benzophenones may allow up to 10 times the exposure before erythema is induced. To provide protection over a wider spectrum, naphthalene-1,5-bisureas have been modified to absorb in both the UV-A and UV-B regions.

Until recently it was assumed that visible light was inert, but it can damage cells and will do so if the flux is high enough or in some cases of photo-dermatoses. Titanium dioxide when dispersed as a powder in a suitable carrying medium such as a cream base can provide good protection through scattering incident light. Colouring agents such as red ferric oxide or a solution of burnt sugar selectively absorb some of the scattered radiation. The use of such a screen either alone or in combination with one of the ultraviolet barriers has been recommended in cases of porphyria, photosensitivity dermatitis/actinic reticuloid, photosensitivity drug reactions, photocontact dermatitis, solar urticaria and summer prurigo (MacLeod & Frain-Bell, 1975).

The use of sunscreens is often frustrated by misinformation or misunderstanding. For example, many fishermen and boat people believe that a lot of solar radiation bounces back off water. But water is completely permeable to ultraviolet light and very little is reflected. Another source of difficulty is that people commonly believe that they do not need protection on a cloudy day. Yet, thin stratus clouds permit the transmission of sunlight. A thin cover of clouds may attenuate only about 20% of the light. There is also the widespread belief that a dark tan protects against the harmful effects of radiation. It is true that the longer wavelengths can produce some tanning without burning, but if the intensity is increased erythema can result. It simply takes a lot more light. Then again it is thought that if exposure is confined to the times when UV-B is largely absorbed by the atmosphere, early morning or late afternoon, then burning will not occur. This is untrue: it simply takes longer.

The use of the term Skin Protective Factor (SPF) in describing sunscreens leads to further misunderstanding. It is commonly thought that white-skinned people should use a screen with a SPF of 6 to avoid burning. The factor simply indicates the relative increase in time to burn. Thus, if a person reddens in 20 minutes without a sunscreen, the use of screen of SPF 2 would allow that person to extend the time to reddening to 40 minutes. Re-application does not allow a further 40 minutes of time since the skin has already received the MED. There is always a permissible dose and this should not be exceeded. The National Institute for Occupational Safety and Health (USA) has agreed standards for maximal permissible 8-hour exposure. The levels vary from 3.0 mJ/cm^2 at 270 nm to over 100 mJ/cm^2 at wavelengths below 200 and above 307 nm (Parrish *et al.*, 1978).

AGEING OF SKIN BY LIGHT

Although alterations in skin with age occur over the whole body, there are differences between exposed and unexposed regions. The exposed skin of the face, hands and arms not only becomes thin, wrinkled, dry and inelastic, but also shows areas of spotty pigmentation and telangiectasia (dilatation of the small arteries and capillaries). These changes are not seen on skin that has been kept covered and there is a clear demarcation. The exposed skin is said to show 'premature ageing' or 'actinic damage' (sunlight-induced damage). Susceptibility is genetically determined, light-skinned Caucasians being most prone while black-skinned people are resistant.

Sunlight-induced dermal changes include a decrease in soluble collagen and an increase in elastin-staining microfibrils and ground substance (Lavker, 1979). There is a continuum of changes in the elastin fibres from curling and fragmentation to simple hyperplasia followed by hypertrophy and finally massive degeneration (Lund & Sommerville, 1957). The papillary region of skin displays more alterations in structure. Changes in fibroblast ultrastructure are most marked. In actinically damaged older skin the bundles of micro-filaments are replaced by tightly packed collagen fibrils. It has been suggested that in 'solar elastosis' the changes in the collagen fibres could account for the appearance of the 'new' elastic tissue (Marshall, 1965). The earliest changes noted were areas of basophilic staining at the edges of the collagen fibres which also took up the elastin stain. Furthermore, there is usually an overall decrease in elasticity in the tissue, so that the best descriptive term might be 'solar elastotic degeneration of collagen'.

Although it may be socially desirable to have a bronzed appearance, the deleterious effect of solar radiation on the skin has to be considered. Tanning is a relatively modern phenomenon. Before this century the rich and socially orientated people of the world avoided tanned skin, prizing white and alabaster. Bronzed, brown or black skin was the mark of the working man. Increased social mobility and more uniform distribution of wealth, sea bathing, and lighter clothes and less of them, all led to great fashion changes. Large areas of skin could be exposed at appropriate times and tanning took over the older status symbol of white skin. But, as Kligman (1978) points out, 'If... it is worthwhile to prevent all this cancer and ageing, we will begin each day by brushing our teeth, putting on antiperspirant, washing our face, applying cologne or perfume and then putting on that sunscreen.'

LASERS

The recent development of lasers has made relatively monochromatic, coherent radiation available in the infra-red, visible and ultraviolet regions. Already lasers are being used in measurement devices, for communication networks and for visual effects in the theatre and open air. Generally the energy densities are well below that which is likely to cause injury, but the development of new applications, particularly in research and medicine, involves high-energy sources which may be a hazard to personnel.

The effects of the interaction of laser radiation with skin and subjacent structures appear to be due to several contributing factors. Among these are the wavelengths of the incident light, the energy dissipation at each site and the conduction rate of heat produced (Klein *et al.*, 1964). The use of fibre optics to conduct the laser radiation to a selected point has enabled tests to be made on the specific effects on different tissue elements (Goldman, Hornby & Long, 1964). It has been found that lasers can be used not only to destroy tissue, but also to weld tissue together. Rockwell's colleagues have used this property to weld Gelfoam packs to the incised edges of liver to reduce the possibility of post-operative bleeding (Rockwell, 1971).

Treatment of various dermatological conditions has been extensively reported; the more successful results have been obtained from tattoos, port wine lesions and seborrheic keratoses. Less successful, so far, have been attempts to treat skin cancer.

TATTOO TREATMENT

The usual method for the removal of tattoos has been continuous-emission ruby and neodymium lasers at energy levels of 50–75 J/cm^2. Argon and YAG systems at power-focussed densities of up to 1500 W/cm^2 have been tried. The Q-switched ruby laser at power output of 4–14 J/cm^2 has also been used successfully, although most reports are confined to tattoo removal from small areas (Rockwell, 1971; Sliney & Wolbarsht, 1981).

Using glass fibre optics with continuous-mode laser radiation small areas of tattoo can be burnt out, but inevitably leave a scar that may at a later date require plastic surgery. This becomes of greater significance when the tattoo is removed from the face or neck or other cosmetically important areas. In many instances, direct removal and transplantation has been the best solution.

Attempts have been made recently to produce a device which would vaporise the dyes without damaging the epidermal cells. A Q-switched laser system developed by M. W. Ferguson-Pell & A. Ritchie (personal communication) uses a focussing system which may be applied directly to the skin so that with a suitable wavelength of radiation the epidermal layers receive significantly lower doses than the dye granules. This device has been used successfully to evaporate tattoo dyes without destroying epidermal tissue (Fig. 51).

PORT WINE LESIONS

Normal continuous-emission ruby lasers at 55–60 J/cm^2 or argon lasers at 5–10 W/cm^2 have been used with some success on port wine lesions to cause collagen fibrotic changes and decreased vascularity (Rockwell, 1971). Thrombogenesis and vessel rupture is followed by healing with reduced red coloration of the tissue. Scarring of the tissue often results and the treatment is effective in only 50% of patients. It is anticipated that superficial vascular port wine lesions may be treated more readily with pulsed ruby Q-switched focussed laser light, but clinical trials have still to prove the effectiveness of this method of therapy.

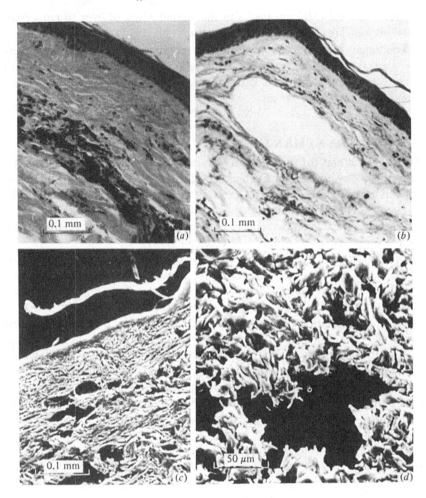

Fig. 51. Laser removal of tattoos. (*a*) Normal skin with dye pigment in dermis; (*b*) after laser treatment all dye has evaporated leaving only a large space indicating original position of dye. The epidermis is largely unaffected by this treatment. (*c*) By scanning electron microscopy no defects in the structure are readily apparent, and at higher magnification (*d*) even the collagen fibres near the site of the laser beam focus are seen to be only slightly damaged. (Photographed from specimens provided by Dr M. W. Ferguson-Pell.)

SEBORRHEIC KERATOSIS

Treatment of common brown spot malady, seborrheic warty keratosis, using a continuous-emission ruby laser at 250–400 J/cm² has been reported as highly satisfactory (Goldman & Rockwell, 1971). With a focussed beam which just covers the small pigmented lesion complete removal and healing

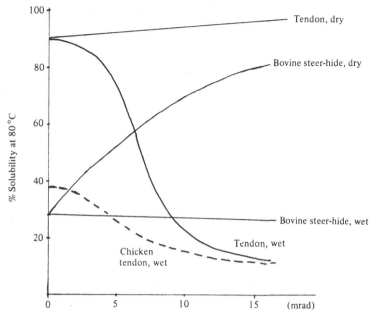

Fig. 52. Effect of ionising radiation on the solubility of wet and dry collagen. (After Bailey, 1968.)

can be effected in 4 to 6 weeks after treatment. Skill is necessary to minimise scarring.

X-RAY AND PARTICLE RADIATION

In the terminology used below, 1 rad refers to a received dose of 100 ergs absorbed energy per gram of tissue and is equivalent to 1 rep or 1 roentgen.

Higher energy radiations (shorter wavelength), including alpha- and beta-particles, all affect the integrity of skin. Basal and squamous cell carcinomas, fibrosarcomas and sebaceous gland neoplasms are known to occur even after single beta-ray doses of 4000–6000 rep. When repeated daily the effect is partially cumulative. A total of about 23 000 rep is needed at a daily dose level of 50 rep to achieve the same effect (Henshaw, Snider & Riley, 1949). Similar changes in tissue can be induced with low-voltage X-rays (Grenz rays) and with normal X-rays generated at 50–100 kV. Effective single doses for rat skin lie between 5000 and 9000 rep. When given over a period of 5 to 8 months at 500–1000 rep per month a total of between 2500 and 7000 rep of 80-kV X-rays is needed to produce similar tissue damage (Zackheim, Krobock & Langs, 1964).

Mast cells appear to be particularly sensitive to this type of radiation. Measurements of histamine levels and mast cell counts after administration of 500 rep of either 12-kV or 30-kV X-rays showed a continuing decrease in both over a period exceeding 50 days (Carlsen & Asboe-Hansen, 1971). The histamine depletion and its duration are known to coincide with inflammation caused by X-rays. Depletion is at its most intense after 24 hours.

Ionising radiation has a direct effect also on the collagen in skin. Whether irradiated wet or dry, 5 mrad is sufficient to cause a significant decrease in the breaking stress of wet tendon collagen. In addition, the same level of radiation induces major changes in the solubility of wet collagen from both skin and tendon (Fig. 52). A review by Bailey (1968) summarises a wide range of effects on collagen. Inflammation, erythema and pigmentation can be readily induced by low-level radiation of skin if this is received over an extensive period of time.

The effects of irradiation on the mechanical properties of skin have been studied mainly in animals. Whether the radiation was given as a single large dose (about 2250 rad) or in a variety of multiple low doses representative of those received in radiotherapy, irradiated skin exhibited a reduced extension at rupture and an increased stiffness. The stress at rupture, skin thickness and load at rupture varied according to the regime of irradiation adopted. Irradiating tissue every third day gave less damage than other regimes and after comparision with clinical data based on a survey of female patients receiving radiotherapy for breast cancer, it was concluded that therapy given three times a week permitted preferential recovery of tissue (Ranu, 1981).

X-rays, beta-rays and gamma-rays all penetrate the skin, only the low-energy radiation being arrested in the dermis. By contrast, all alpha-particles are absorbed by skin, a large proportion in the epidermis. However, penetration of the vehicle carrying the radioactive element may occur through the hair follicle, sweat ducts and glands. It follows that as far as skin is concerned the effect of gamma-radiation is negligible compared with that from alpha-particles.

Witten, Sulzberger & Wood (1957) demonstrated that alpha-particles from polonium-210 could produce an erythemal response if given in high enough dose. The threshold dose is approximately 75000 rad and the onset of erythema is largely a function of total dose received. The higher the dose the sooner was the onset of erythema, which appeared in 2–6 hours and reached a maximum in 12–18 hours. Thereafter, it gradually subsided. Pigmentation of skin is noted in 36–48 hours and by 72 hours may obscure the erythemic response. The pigmentation produced by the radiation may last from 10 days to several months. Where damage to the skin is great it peels in a manner similar to that of sunburn.

The fact that thorium X, an alpha-emitter, has been used in dermatological therapy for more than 60 years without producing any serious sequelae and that polonium-210 appears to produce the same responses, leads to the suggestion that alpha-particles are responsible only for the erythemic response and pigmentation whereas both beta- and gamma-radiations produce not only atrophy, epilation and telangiectasis but also chronic ulcers, keratoses and skin carcinomas. The nature of the changes that occur in tissue as a result of radiation is still not understood. Among the changes which have to be considered are the production of secondary radiation and formation of new chemical substances as well as direct physical damage to blood vessels of the papillary dermis.

7

WOUND CLOSURE AND ALTERNATIVES
TO SKIN

The closure or treatment of a wound so that it heals and resists rupture is a primary consideration. Whether the wound be caused by incision, avulsion, ulcer or burn, the integrity of the body surface has to be restored. In all but superficial wounds, the deposition and orientation of new dermal collagen, the restoration of an appropriate vasculature and the re-establishment of the epidermis have to mimic the original tissue unless unsightly scarring is to ensue.

Since the time of Hippocrates, whose writings indicate some understanding of the problem, it has slowly become established that dermal connective tissue can alter in response to stress and hence affect the way in which a wound heals. In 1926, Bunting and Eades found that mechanical tension could orientate fibroblasts in fresh wounds in the direction of applied tension. Even the mitotic spindles within cells were aligned parallel to applied stress and daughter cells separated along the lines of tension.

Many reports note in wounds healing under tension that newly deposited collagen becomes orientated in the direction of stress (Brunius & Ahren, 1969; Arem & Madden, 1976). Rupture or even disruption of the newly formed collagen will delay healing. Other forms of stimuli such as the physiological stressors, increased adrenocorticotrophic and medullary hormone output and changes in protein, glucose and lipid metabolism are all thought to retard healing since they are known to affect maintenance, growth and repair of skin (Cohen, 1979).

On extensive wounds it is necessary to assist regrowth of epidermal tissue by means of grafting. Frequently the wound bed is unsuitable for the support of such a graft. When the deep vasculature is damaged it may not be possible to provide an adequate supply of nutrients to the graft tissue and surgeons have learnt to wait until a supply is restored, as indicated by restoration of granulating tissue, or by direct connection of blood vessels to the graft. Complications arise with infection of the site and hence a variety of procedures have been evolved to deal with the different types of wounds encountered. In the following sections we shall be concerned only with the closure of simple wounds and extensive wounds caused by ulceration or burning.

SUTURES

To prevent an incision type wound from re-opening under tension, many systems for the apposition of the wound edges have been used, the most common being suturing and tape closure. The purpose of a suture is to coapt the tissue until the wound gains adequate strength, but each insertion of the needle inflicts further trauma and many suture materials produce tissue reaction. Over-tightening the suture crushes the enclosed tissues and jeopardises the blood supply. In practice it has been found better to use small needles with fine smooth sutures placed closely together with minimum tension. All deep wounds are best treated with buried sutures with the knot placed deeply in the dermis and final closure using a simple superficial running 'blanket' stitch (McGregor, 1965).

The ideal suture would be one which held until the strength of the wound was equal to that of normal tissue and then was promptly and completely absorbed. As yet such sutures are not available. Many of the new synthetic materials have advantages over the traditional silk, plain catgut, chromic catgut and collagen, but, while flexibility and uniformity of handling can be provided through production control, some synthetic materials do have poor knot-holding characteristics. Polypropylene (Prolene) monofilament nylon and stainless steel do function sufficiently long, but are non-absorbable. The absorbable sutures, polyglactin 910 (Vicryl) and polyglycolic acid (Dexon) both cause minimal tissue reaction, but lose strength rapidly.

Polyglactin 910 loses about 60% of its tensile strength by the 18th day and is completely absorbed within 90 days. Polyglycolic acid loses 80% of its tensile strength in 14 days and is absorbed in 90 days. By comparison, collagen fibres inserted into rat skin can still be identified 12 months after insertion and silk and catgut are enclosed by connective tissue as a foreign body. On the other hand, the strength of the healing wound may be higher with absorbable sutures though differences are not usually found during the first 21 to 28 days post-operative period. By 70 days the differences are quite marked, but since the incidence of wound infection is higher with non-adsorbable materials, part of these differences may be due to the effects of infection rather than an inhibition of growth by the presence of the suture.

The strength of the wound depends upon the healing of the dermis. The organisation of the collagen fibrils does not begin until days 4 to 6 and the wound achieves only 20–30% of normal strength by the 14th to 21st day, and 60% by the 120th day. Where wounds fail to heal, as in bed-sores, the unorganised collagen bundles remain a feature of the dermal region (Fig. 53). During the early phase of healing there is a characteristic inflammatory reaction with removal of debris, deposition of non-collagenous proteins and glycoproteins and the migration of fibroblasts and epithelial cells. The wound is held together by a fibrin glue which may form a hard scab at the surface where it dries out. As the collagen bundles begin to form, there is a rapid

Fig. 53. Transmission electron micrograph of collagen bundles in bed-sore. There is no apparent direction of orientation for these relatively small bundles and the tissue is extremely fragile.

increase in the strength of the wound. Tensile strength is increased as the collagen fibres become more insoluble and are remodelled by interweaving, closer packing and reorientation.

Monofilament sutures are superior to multifilament types, particularly with

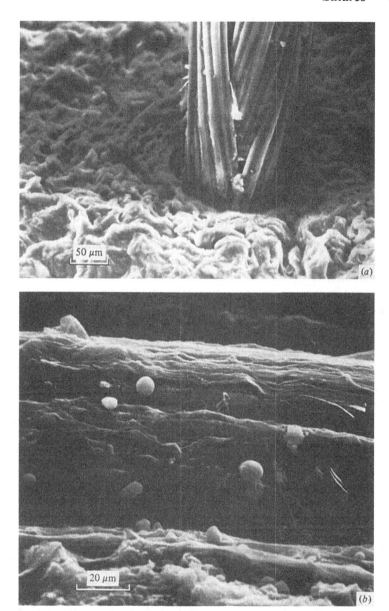

Fig. 54. Sutures in skin. (*a*) Woven synthetic fibre in dermal collagen. Cells and intercellular substances remain attached wherever the fibres of the suture cross. (*b*) Collagen suture in dermis. Even after a week, this suture remained free and was not penetrated by host tissue.

regard to the incidence of wound infection (Van Winkle *et al.*, 1975). By scanning electron microscopy it is possible to identify cells attached to multifilament sutures of all types of material, which on close inspection seem to have an affinity for the fibre where the distance between adjacent fibres is approximately the same diameter as the cell. This is seen frequently in the case of twisted silk sutures. In all cases the suture is enclosed by connective tissue, is often invaded, and frequently uncoils to permit movement of fluid along the suture. Intense granulocyte reaction is seen in surrounding tissue, particularly in the case of silk. The absorbable sutures tend to be more tightly associated with connective tissue, with macrophages and fibroblasts invading the suture. Some examples of suture material in skin are given in Fig. 54.

Silk and Mersilene generally give rise to the highest rate of wound infection and the most intense tissue reaction. Prolene and other monofilament sutures produce only mild reactions. When dealing with contaminated wounds, however, the egress of fluids and the avoidance of additional trauma are important. Polyglactin 910, polyglycolic acid or nylon are suitable closure materials after lavage and adequate surgical debridement (Stevenson, 1977). When used in general surgery, polyglycolic acid sutures are at least as good as silk for sub-cuticular closures and are more convenient to use (Clough & Alexander-Williams, 1975).

TAPE CLOSURES

One of the earliest methods of closing skin wounds was to glue tape on each side of the wound and then suture the tapes rather than the skin. A more sophisticated version of this method was introduced by Dioguardi & Musajo-Somma (1977) who prepared four pieces of adhesive tape linked together with two continuous sutures, 1 to 1a and 2 to 2a. The two pieces of tape were laid on either side of the wound and the wound closed by applying tension to the complementary tape; the upper tapes were then stuck down. The edges of the wound could be easily adjusted between the threads to ensure that they lay in close apposition (Fig. 55). Advantages were said to be speed of application and unhampered observation of the wound at all times. It was claimed to be especially suitable for children.

The use of adhesive tape to close a wound directly has been advocated since such dressings were available in sterilised packages. Advantages appear to be many: (*a*) greater tensile strength of the wound, (*b*) improved protection against infection, (*c*) better cosmetic result and (*d*) improved remodelling of collagen (Forrester *et al.*, 1970). In test animals (rats) the tensile strength of tape-closed wounds was about 27% higher than that of suture-closed wounds. After 150 days the tensile strength of tape-closed wound was only 12% less than unwounded skin whereas sutured skin was 31% less.

In a comparative study of sterile 'Scotch' tape and Johnson and Johnson microporous tape with 2–0 silk suture on wounds inflicted on the backs of

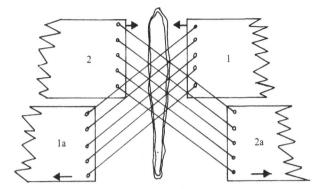

Fig. 55. Wound closure by tapes. Long slit wounds may be closed by drawing together the two tapes (1 and 2) placed alongside the wound, which are then held in place by pressing the self-adhesive top sections (1a and 2a) to the skin. The wound remains open to the air to aid drying.

pigs, Carpendale (1964) showed that surgical wound infection could be markedly reduced. Even inoculating the wounds with *Staphylococcus aureus* before closure failed to produce an infected wound when tape-closed. The absence of suture marks also leads to a better cosmetic result since these often remain as white dots for many years after healing. Moreover, the effect of dressing pressure is thought to be helpful and the elasticity in the adhesive dressing can be used to provide a small pressure across the wound area.

Ordinary adhesive tape has been found to be beneficial in healing more extensive wounds, such as leprous ulcerations. The tape inhibits the growth of certain bacteria, especially Gram-positive staphylococci and streptococci and has, therefore, been advocated for the treatment of deep burns, for the dissolution and removal of necroses, and for cleaning out greasy and infected wounds (Stenstrom, Bergman & Bergman, 1972). Stenstrom *et al.* also used wide adhesive tape for direct application to large burn areas. The adhesive does not stick to the wound and can be removed each day, generally without the need for local anaesthetic. When used with a covering of absorbable material it aids the drying of suppurating regions.

TISSUE ADHESIVES

Despite the fact that tissue adhesives have been available since 1962, there is still uncertainty about their use. This may be due to reports of tissue toxicity, especially with methyl 2-cyanoacrylate (Raekallio & Seligman, 1964). However, the longer-chain homologues such as butyl 2-cyanoacrylate are much less toxic and have been used successfully in closing skin wounds (Alhopuro *et al.*, 1976). Wounds closed with adhesive have much less epidermal tissue reaction than those closed with sutures. However, closure is not faster than with sutures and it is impossible to obtain satisfactory cosmetic results in wounds extending through the dermis unless adhesive is used in conjunction with sub-dermal sutures or widely spaced dermal sutures. The adhesive is then used for final tissue alignment and support. Poor technique has been found to be the major cause of unsatisfactory scars. It is important that the edges of the wound are brought into close proximity before the adhesive is placed on the skin surface. Any adhesive that penetrates between the epidermal margins delays healing.

With good technique the method has value in closing wounds in children. Pain-free immobilisation of skin margins can be achieved without anaesthesia and since skin in children heals faster, the adhesive can be left to self-release through cell desquamation as the wound heals. Although it is possible to obtain cosmetically better results than those obtained by suturing, satisfactory scarring is only achieved when no tension is exerted on the sealed edges. This may be accomplished when adhesive is used in conjunction with adhesive tape as dressing.

TREATMENT OF BURNS

From earliest records it appears that man has attempted to treat burns either by unguents or by covering the site with a suitable dressing. The Ebers papyrus tells us that milk, gum and goat's hair were laid over the burned region of skin. In the Smith papyrus, other commentators have reported the use of linen strips soaked in oil. The Chinese used tea-leaves as a dressing, while Celsus prescribed a mixture of bran and honey. Since that time there has been an almost bewildering number of recommended treatments and still the search continues to find a suitable covering for burns and a replacement for skin.

One effect of heating skin is to destroy the semipermeable membrane associated with the lipoprotein in stratum corneum. Breakdown of this layer allows water loss through the damaged skin and since the heat of evaporation for water is about 2.43 MJ/l, a substantial evaporative water loss also means an abundant loss of heat. Heat loss can be reduced by surrounding the patient with warm air, but it would be rational to replace the lost lipoprotein layer with a fatty layer, an idea dating back some thousands of years. The control of evaporation from burns patients, then, is one of the primary objects of initial treatment.

To be able to compare different methods for reducing evaporation loss, adequate techniques have to be available to measure actual water loss from the body surface. The testing of materials *in vitro* is relatively easy and there are established standards both for measurements and for the method of expressing the values obtained (ASTM Standards, E96–66, 1972). For measurement *in vivo* some form of evaporimeter is utilised. This instrument depends upon the physical law of water diffusion into air from a surface (see also Chapter 1). Close to the surface, within about 1 cm, the water transportation is determined by the formula

$$\frac{1}{A}\frac{dM}{dt} = -D\frac{dp}{dx}$$

where A is the area of the surface (m²), M the mass of transported water (g), t the time (h), p the partial pressure of the vapour in air (mmHg or Pa), x the distance from the surface (m), and D a constant $(0.67 \times 10^{-3}$ g·m⁻¹·h⁻¹·Pa⁻¹ or 0.0877 g·m⁻¹·h⁻¹·(mmHg)⁻¹.

Thus, the evaporation rate dM/dt is proportional to the partial pressure gradient dp/dx. A variety of instruments is now available using this principle (e.g. Servo Med AB, Sweden).

Using an evaporimeter, Lamke & Liljedahl (1971) measured evaporative water loss from burns, grafts and donor sites in a slow current of air (2.5 cm/s) over a site of 25 cm², recording the increase in humidity. From values obtained they calculated evaporation. Their equipment was more basic than the current commercial products; nevertheless, it produced data which have been the basis of most research since that time. They showed that first-degree burns (epidermal damage only) retained water almost as well as normal skin. Second-degree burns (heat damage to either the superficial dermal layer or to both superficial and deeper dermal layers) exhibited massive water losses some 24 times greater than the rate from normal skin. Third-degree burns (destruction of epidermis and dermal layers) gave off substantially less water. Maximum water loss for third-degree burns has been shown to be about 15 times normal (Lamke *et al.*, 1977*a*). Although both types of second-degree burn lose massive amounts of water, there is a substantial difference between the two types. Superficial second-degree burns begin to lose water at the same rate as deep burns, but the rate quickly subsides even within a few days. By comparison, deep second-degree burns maintain high water loss for 10 days or more and do not return to normal levels even after 20 or more days.

Protection of the burn by dressings may have consequences that appear to be contrary to the desired effects. While occlusive and semipermeable dressings may arrest or slow down evaporation, they also lead to a decreased mitotic rate in epidermal cells and hence slow down the rate of healing (Fisher & Maibach, 1972). Generally these dressings are tissue based and are known as grafts even though there is no intended incorporation of any part of the dressing into the tissue of the healing wound. The term *autograft* is used for transfer of tissue from one site to another on the same patient and *homograft* is used synonymously with *allograft* for transfers of tissue from the same species. The terms *heterograft*, *xenograft* or *zoograft* are used to describe tissue transfer from a species other than the recipient. All grafts are also commonly referred to as membranes.

Homograft membranes, whether split-skin or whole skin, slow down the rate of water loss with permeability *in vivo* becoming similar to that measured for the materials *in vitro* (Lamke *et al.*, 1977*b*). The autograft gives no early reduction in total rate of evaporative water loss since the autograft donor sites are initially freely water-permeable. Indeed, post-operative water loss can be higher than that measured pre-operatively. This is why donor sites should be covered as soon as possible. Heat loss, in the case of third-degree burns, has been estimated by a number of workers and, when the combined loss due to radiation, convection and evaporation is evaluated, it may total 3500–4500 cal/kg per day. Of this some 45 to 80% is due to evaporative water loss (Moncrief & Mason, 1962).

NATURAL MEMBRANES
HOMOGRAFTS

Most of the early work was concerned with attempts to transfer heterografts such as the skin of frogs, lizards, rabbits, kittens, pups and even birds. Commonly, the process was accompanied by incantation and witchcraft. It is only since the late nineteenth century that successful grafts have been reported. The first autografts were applied in the 1870s and the first successful homograft was reported in 1881 (Rogers, 1951). As surgical techniques improved so homografts became invaluable for the closure of open granulating wounds. Once the homograft has been applied pain disappears, evaporation decreases, exudative protein loss is minimised and bacterial growth discouraged.

Homografts have been used with increasing frequency as temporary dressings. They are normally removed each day and replaced by fresh grafts until the tissue becomes adherent. The wound is then ready for final grafting and in defects involving full-thickness skin this technique has proved most valuable. The temporary dressing when used as a test for suitability for permanent grafting has also been advocated for any extensive wound however inflicted, contaminated wounds, and wounds which would otherwise be difficult to manage (Shuck, Bedeau & Thomas, 1972).

The main difficulties with homografts are preparation and storage. Viable (fresh, untreated) skin is difficult to collect and transport but is cheap, available anywhere and particularly useful for full-thickness defects. Tissue-typed frozen skin is more expensive to prepare and store, but is available when required in large surgical units. It must be taken within a few hours of death and prepared carefully. To sustain its condition, the tissue is normally placed in 'glycerolised water' containing 15 ml glycerin, 2.5 ml sterile sodium bicarbonate solution (4.4%) and 97.5 ml distilled water. After about 30 minutes it is transferred to nylon film bags containing 2 ml 'Transporting Solution' consisting of 2.5 ml sodium bicarbonate solution (4.4%) in 87.5 ml distilled water with 10 ml '199 medium' concentrated 10-fold. To this is added 250 mg ampicillin and 250 mg cloxacillin (Cochrane, 1968).

A number of these bags are then placed in the freezing chamber of a

Bio-freeze machine, which gradually reduces the temperature from 4 °C down to −30 °C at a rate of 1 deg C per minute. The freezing process is then automatically accelerated until a temperature of −70 °C is reached; after this the process is continued at a rate of 10 deg C per minute to a final temperature of −100 °C. The bags, suspended on frames, can be stored for up to 2 years at the temperature of liquid nitrogen. Reconstitution is achieved by immersing in a water bath at 37 °C for approximately 1 minute.

Radiation-sterilised skin has been used on granulating burn wounds as split-skin grafts obtained from human cadavers. For radiation sterilisation the total amount of energy needed is about 3–4 megarads (1 rad = 100 ergs/g). A cobalt–60 source developed for clinical use and containing about 4000 curie (Ci) is often used for this purpose. After irradiation the sterilised homografts are stored at −20 °C to −25 °C (Korlof *et al.*, 1972).

Radiation-sterilised split skin reduces evaporation by about 85% but rapidly loses its effectiveness after 3 days. If applied to receptive skin, however, developing epithelium begins to take over the barrier function. The sterilised skin is rapidly phagocytosed by the host and after 9 days there are only a few dead cell layers left of the original dressing.

A third way to prepare skin as a dressing is by freeze-drying. The final product is known as lyophilised skin. When reconstituted in normal saline it serves the same purpose as fresh or frozen skin and is thought to be a superior dressing. When grafted onto receptive surfaces it will form a true graft (Hackett, 1975). For preparation, skin can be collected up to 3 days after death and stored in Transporting Solution at 4 °C pending lyophilisation. The skin is first frozen to −40 °C and then transferred to a freeze drier, such as the ChemLab SB 5 Freeze-Drier. The dry skin is sealed in polythene bags and sterilised with ethylene oxide. Storage is unrestricted. When prepared in this way skin can even be sent by post (unlike fresh or frozen skin, where transport is a problem), which means that this type of preparation can be available at all clinics where burns are likely to be treated.

Assessment of lyophilised skin as an initial dressing in partial thickness burns has been reported by McDowall & Hackett (1976), who concluded that it ensures optimum conditions for healing through mechanical protection of the burn area, reduction of infection and prevention of evaporative water loss. When used on the hand, the resulting relief from pain helps early mobilisation of the joints and early dispersal of oedema.

Chorion and amnion

About 7 to 8 days after fertilisation of the human ovum, there is a separation from the inner cell mass of the germ disc at the periphery of the ectodermal layer to form the slit-like amniotic cavity. The amniotic epithelium becomes separated from the primitive trophoblast when the primary extra-embryonic mesoderm begins to form. The epithelium contains many mitotic figures and the cells rapidly become flattened as the pregnancy progresses towards term. The amniotic membrane, therefore, has two components, one derived from the ectoderm and one from the extraembryonic mesoderm that forms the chorion. At about the second month of gestation the mesenchymal cells become separated from the epithelium by a connective tissue layer of loosely packed collagen fibrils interspersed by active fibroblasts so that, at term, the foetal membrane consists of a single layer of epithelial cells (of ectodermal origin) firmly attached to a collagen-rich mesenchyme which is itself loosely attached to the chorion.

The first reported use of foetal membranes in skin transplantation was given by Davis (1910). Since then amniotic membrane has been used successfully in the treatment of burns and ulcerated surfaces. At first, the membranes were applied directly to the wound, covered with warm paraffin and then with dressings. After 48 hours the dressings were removed, when it was found that the amnion adhered to the wound and the chorion to the paraffin. However, the technique was not adopted generally and it was some 30 years later that further interest began to develop.

Kubanyi (1945) used live amnion in patients with burns and traumatic skin wounds including the repair of a high small bowel fistula. From then on a number of experimental procedures were reported embracing well documented research reviews (Douglas, 1952; Sterling, 1956; Pigeon 1960; Pringle, 1963). But, in a review of methods of resurfacing denuded skin areas, Pruitt & Silverstein, in 1971, made only a passing reference to the technique and that in a dismissive form.

Robson & Krizek (1974) and Trelford, Hanson & Anderson (1975) eventually acquired sufficient experience and reputation to establish their techniques in surgery. In particular, their application of the method to non-healing wounds in diabetic patients, as a graft over the surgical defect of total glossectomy and as a biological dressing to cover the entire peritoneal cavity left bare following debridement for necrotising cellulitis caused by gas gangrene has been acclaimed. A review by Trelford & Trelford-Sauder (1979) of the use of amnion in surgery outlines the current practical applications in medicine.

The strength of foetal membranes was tested originally by dropping cannon-balls onto the membrane stretched over a ring. More recently, MacLachlan (1965) measured the bursting pressure, and Polishuk, Kohane & Hadar (1964) measured the load at rupture to be 0.05–0.45 kg/cm width of membrane. It appears that smaller foetuses have stronger membranes, since tensile strength diminished with increasing foetal weight.

Water loss through foetal membrane has been measured by Lamke (1971), who found that direct application to a second-degree burn or third-degree site reduced water loss by only 15% compared with 91% by a homograft. A similar result was found on granulating wound sites. When used on fresh wound sites – donor sites – reduction was slightly less than 20%. This would indicate that amniotic membrane can give good protection of a wound site but requires an occlusive dressing to be effective in the treatment of burns. Favourable results in the treatment of ulcers were obtained with split-thickness preparations of the single-cell-layered amnion and its basement membrane, followed by autografting 5 days after commencement of treatment (Bennett, Mathews & Faulk, 1980).

Meshgraft

By introducing a regular series of slits into the graft tissue it can be expanded to form a mesh in much the same way as a sheet of metal can be expanded. This is used principally with split-skin grafts for covering large wound areas in both burns and donor sites. The principle of the technique is to provide viable skin in a regular pattern over the area and so 'seed' the attachment and regrowth. It is an effective way to increase rate of healing but, as Lamke (1971) showed, the mesh is no better, though no worse, than foetal membrane in preventing water loss. Additional protection against excessive water loss has to be provided.

HETEROGRAFTS

The first scientific understanding of foreign skin grafts was provided by Silvetti *et al.* (1957), who demonstrated their value as a temporary biological dressing. The concept of a biological dressing is based on the fact that there is no vascularisation of the applied tissue or any sign of cell-mediated rejection. Tissue from animal origin can serve only as a dressing, but two immediate gains are obtained in the treatment of burns and similar wounds: (*a*) cessation of pain and (*b*) reduction in water loss. In properly prepared wounds the heterograft will also adhere to the wound surface.

Heterografts can be made available at all times either as fresh, frozen or lyophilised dressings. However, the commonly used pig skin heterograft is characterised by early rejection and abundant seropurulent discharge. Silvetti and his colleagues (1979) advocate the use of bovine heterografts, claiming that they exert rapid bacteriostatic and bacteriocidal effects, stimulate formation of type III collagen and have pronounced angioblastic activity. Bovine skin quick-frozen in liquid nitrogen can be made available in most areas of the world.

The use of lyophilised pig skin has been advocated by many groups including Wood & Hale (1972), Elliot & Hoehn (1973), Chatterjee (1978), Morris, Hall & Elias (1979) and Ersek & Mayer (1980). There is a lower incidence of infection compared with open treatment or cotton dressings, with bacterial counts reported to be reduced by as much as 83%. In the lyophilised form the dressing is available commercially; Armoderm (Armour Pharmaceutical Co., Sussex) is generally obtainable. The 'skin' consists of both dermal and epidermal layers, with a total thickness between 0.2 to 0.38 mm, available in five different sizes. It is sterilised by gamma radiation and has a shelf life of 2 years.

When used with antibiotics the incidence of infection is very low. Healing time is rapid (6–20 days depending on size of injury and depth of burn), and reduction in evaporation rate is better than 83% for partial-thickness (second-degree) burns, full-thickness (third-degree) granulating burns and donor sites (Lamke, 1971). The time to healing is about half that obtained with paraffin-impregnated gauze dressings (Chatterjee, 1978). Lyophilised skin has also been reported as suppressing hypergranulation tissue (Roberts, 1976).

Modified collagen

The use of an artificial skin as a dressing or graft has long been the goal of those who see natural membranes as having too short a storage life or too costly a preparation. Because of the many difficulties encountered in the preparation of a synthetic dressing, and in particular with biocompatibility, attempts have been made to obtain a membrane based upon collagen. As a result of improved methods of extraction and purification exogenous collagen has been used in many forms of dressing. The advantages of collagen are principally that it can be isolated and purified in large amounts, its antigenicity can be altered, and it can be made into a variety of physical structures.

Collagen has been used in many forms: as reconstituted fibrils, reconstituted extruded strips, reconstituted sheets on Dacron mesh, collagen-fabric films, microcrystalline porous mats, microcrystalline sheets, microcrystalline powder, and as tanned collagen sponge grafts. In the microcrystalline form it has been shown to be intensely thrombogenic.

Collagen mats or 'sponges' have been advocated as dressings to provide a temporary fibrous template for new connective tissue. The antigenic activity of collagen is small and claims have been made for its universal application. In the membrane form, collagen dressings are said to reduce time to grafting by up to 10 days (Tavis *et al.*, 1975). It provides as good a protection of the surface as natural dressings and reduces water loss by about the same amount as heterografts. Bacteriological counts indicate a good suppression of infection, with values similar to those with autografts.

All collagen preparations have a tendency to dry out. To prevent this a variety of techniques have been employed including that of cementing a plastic film, such as polyvinyl chloride foil, over the mat or sponge. Such a collagen foil dressing has been marketed by a German manufacturer; it is many millimetres thick. Another version uses a polyurethane foam in a special bandage. Foams can be made from dried, fine fluffy fibres (such as those marketed under the trade name Avitene) by first hydrating in a mixture of water and petroleum ether to prevent complete cross-linking and, if not in acid form, swelling by addition of 0.8% lactic acid. The resulting gel can then be foamed by a variety of different agents and stabilised (tanned) by means of an aldehyde such as glutaraldehyde. The foam is generally neutralised by adding sodium bicarbonate and dried to give a porous collagen film. The addition of a plastic film to control transmission of water is then recommended (McKnight & Guldalian, 1974).

Dressings of the foamy type are soft, pliable, flexible and so conform to wound topography. When applied under test conditions to animal skin, healing was observed over a 4–7-day period. Clinical evaluations of such dressings show that they are no better than homografts or heterografts and tend to be more expensive.

ARTIFICIAL FILMS, AND FOAMS AND LAMINATES

FILMS

Monolayer films and sprays form probably the largest group of synthetic dressings for burns. Many of the early attempts using Aeroplast, Cellophane, silicone and thermoplastic vinyl copolymers were unsuccessful due to poor adherence to the wound. Suppuration from the invariably infected burn wound meant that a superstructure of pressure dressings was required to support the film.

Silicone rubber, being an inert material, was an obvious early choice as a possible dressing material but, although the application site was free of pain, wounds were slow to heal and displayed heavy *Pseudomonas* growth. Even the sponge-like membrane formed from formalised polyvinyl alcohol (Ivalon), while allowing good fibroblastic invasion also provided an excellent environment for bacteria. Sub-membrane purulence and hardening of the membrane detracted from the usefulness of this product.

Graft-take was improved by surgical debridement of wounds. Velours of nylon 6, Dacron and rayon were then found to have greatly improved adherence, but still inflammatory reaction was moderate and infection and foreign body reactions remained as contra-indications for their use. It is not surprising, therefore, to find in reviews of that time a somewhat despondent attitude and pleas for more development work (Pruitt & Silverstein, 1971).

Polyurethane, polylactic acid and polycaprolactone have all been used with some degree of success. Of these, polycaprolactone has the most appropriate physical properties for wound dressings and experience with animal tests confirmed the proposition. Silastic membranes, however, continued to be used even though silastic was already well understood to be a completely impermeable membrane and to retard healing (Harris, Filarski & Hector, 1973).

A number of manufacturing companies are now producing a variety of different films, one of which, Opsite, an elastomeric polyurethane (Smith and

Nephew Ltd), had temporary approbation in a number of clinics as a useful, readily available dressing. It is permeable to water vapour but not to bacteria and healing is faster. The disadvantages were that fluid leaked from under the dressing and newly formed epithelium tended to adhere to the film so that removal of the dressing took with it the new tissue, leaving behind a raw surface that was slow to heal. The necessity to aspirate the wound during the first 48 hours is an inconvenience and the addition of a padded dressing to absorb the fluid has been advocated (James & Watson, 1975).

Polyvinyl chloride (PVC) and polyethylene films are both produced in large quantities for wrapping and presentation of food. In the treatment of hand burns, the advantages of a plasticised polythene bag compared with conventional dressings are manifold. The technique is simple, painless, more economical of nursing time and very cheap. It has proved a useful technique when combined with antibacterial cream inside the 'glove' to control the growth of sepsis, and allows stiffness and oedema to be minimised by exercise within the bag (Reid, 1974). Even larger bags to enclose the proximal limbs have been advocated by Lendrum & Bowen-Jones (1975), so effective were their results.

Very thin plasticised PVC sheeting sold as Perfa Cling (Perfa Rap Ltd, High Wycombe) has been used as a wound cover (Lendrum & Bowen-Jones, 1975). The procedure was simple: the wound and adjacent undamaged skin were cleaned carefully and any dead tissue removed with scissors and forceps. A single sheet of PVC was placed over the wound and attached at the edges by adhesive tape. Orthopaedic wool was placed at the free edges of the dressing to mop up any fluid exudate. The dressing was changed every day, the dressing peeling away readily and painlessly. The only disadvantages reported were related to the development of rashes on adjacent normal skin. This was minimised by careful prewashing of the area.

In the treatment of donor sites Clingfilm (Vitafilm), made by Goodyear Tyre and Rubber Co. Ltd, has been recommended as an alternative source of plasticised PVC. As Vitafilm it can be obtained in a variety of widths and thicknesses, is elastic and flexible and sticks readily to itself. Strips of Vitafilm can be autoclaved in order to sterilise the film and if protected by autoclave crêpe paper the shrinkage remains small (about 5%). Townsend (1977) used the adhesive Mastisol to attach the film to the dry donor site, time having been allowed for clotting to occur. The film remained in place throughout healing.

Although PVC is known to have very low water vapour transmission the thin commercial films have adequate permeability. Vitafilm, grade F10, which is less than half the thickness of Opsite, has a water vapour transmission also of half the value for Opsite: about 800 g/m^2 per 24 hours. Its oxygen permeability, however, is significantly greater and this is important in wound healing.

Films suitable for dressings can be made from polyamides such as polyethylene terephthalate as well as polyvinyl acetate and vinyl acetate/vinyl chloride copolymers. These can be made as microporous or perforated sheets to allow transmission of water. Bearn & Barsby (1977) suggested that these sheets could be printed by any convenient method to represent skin. A similar inert polymer is polytetrafluoroethylene (PTFE). An electrostatically induced negative charge reduces adherence and intimate contact with the wound tissue, but despite this, laboratory trials indicate that PTFE might be used effectively as a wound dressing.

FOAMS

Because of the difficulties encountered with adhesion of films to burns, thicker, textured materials were thought to offer advantages, particularly in encouraging tissue ingrowth. One of the earliest attempts to develop this idea utilised a formalinised polyvinyl alcohol (Ivalon) (Chardack, *et al.*, 1961). It adhered to a denuded body surface through growth of tissue into the pores and, by virtue of its hydrophilic character, water loss could be controlled. Disadvantages soon appeared when biopsies taken from underlying tissue after removal of the dressing showed embedded fragments of Ivalon impurities. If allowed to dry out, Ivalon became rigid. It currently finds no favour as a dressing.

Another foam material, Epigard (Parke Davis and Co.), is composed of an inner layer of reticulated polyurethane foam which has been laminated to an outer sheath of microporous polypropylene film. Its beneficial effects appear to be most closely related to coagulation of the plasma exudate and provision of an interface for bacterial−phagocyte interaction. There is, therefore, a need to keep it in close contact with the burned surface (Alexander *et al.*, 1973). Use of Epigard is contra-indicated over dead eschar before the stage of separation, but it is recommended to debride wounds at the stage of eschar separation and to accelerate the development of healthy granulating surfaces prior to grafting.

An alternative to Epigard has been patented by VEB Synthesewerk Schwarheide who bonded a polyurethane foam to a polyurethane film made by irreversibly compressing a sheet of foam, to provide a controlled gas and water-vapour permeable membrane (Foritz *et al.*, 1979). So far, we are unaware of any clinical data to demonstrate or confirm the value of this material.

In an attempt to provide an ideal dressing for a granulating wound, a polymerised silicone foam elastomer has been tested by Wood, Williams & Hughes (1977) and found to be non-adherent, non-allergenic and slightly absorbent. The material is Dow Corning silicone (X7–9100) with catalyst. After mixing appropriate quantities of the silicone and catalyst it is delivered directly into the wound where it expands to about 4 times its own volume, filling the wound and setting to a soft foam. The foam can be held in place by Micropore tape. The authors suggest that the material is only an improved method for packing a deep wound, but their clinical experience suggests that healing is faster. This experience, from 250 patients, has been supported by Harding & Richardson (1979) who suggest that the technique may be extended from special hospital clinics to routine use by district nurses.

LAMINATES

With developing interest in textured materials for burn and granulating wound dressings, fabrics seem to be an alternative to foams. The lamination of semipermeable membranes of silastic, polyvinyl chloride (Saran wrap), polyurethane and polymer films to the velours of nylon, Dacron and rayon was tried initially.

When a semipermeable barrier was added to control water loss, it also stiffened the membrane so depriving it of its necessary flexibility and elasticity. Attempts to improve the system were reported by Kornberg *et al.* (1972) who bonded ultrathin membranes of silicone to spun-bonded nylon (Cerex), open-weave nylon and double-knit Dacron. Animal tests gave excellent results and indicated its potential in the treatment of full-thickness skin defect. On patients, results were at least as good as those from heterografts in covering second-degree burns and donor sites.

These composite materials probably approach nearer to an artificial skin than a film or simple foam dressing. The term 'skin', however, implies that the material might take over the properties of natural skin and, so far, this is far from being the case. Even laminates with collagen, though performing well as a dressing, are no substitute for skin.

By contrast some laminates have been developed simply as a dressing and one of particular note is that described by Nathan *et al.* (1976), who claimed that their unique system protected severe wounds such as second- and third-degree burns from infection by isolating the tissue from microbiological contaminants arising in the patient or the environment. Experiments *in vitro* showed that the films produced *in situ* depressed bacterial growth in both aerobic and anaerobic conditions, that bacteria had poor survival on the film and also that antibiotics could penetrate the material.

When used on wounds the film is built up from two components: a powder (PHEMA) and a solvent (Polyethylene glycol 400, PEG). The combined film PHEMA–PEG can be built up in layers. The PEG is spread over the burn surface from a syringe and evenly dispersed with a swab. The powdered PHEMA is applied from a container with a perforated top so that the polymer is dusted onto the PEG. The mixture takes about 30 minutes to form a solid film.

This dressing has a number of advantages which appear to be upheld in clinical experience: (*a*) the components are suitable for direct tissue contact showing no toxic response, (*b*) need for dressing change is significantly reduced, (c) control of any bacterial infection can be achieved through topically applied agents. Because the dressing is formed directly on the wound surface, the chance of spaces arising between the dressing and the wound surface is minimised. Incidentally, when the material is used as a wound dressing, it is called the Hydron Burn Dressing.

Although the PHEMA-type systems have been claimed to be novel, the Pluronic range of copolymers could serve equally well. Pluronics are commercially available high molecular weight block copolymers derived from a propylene glycol initiator. They are made by the sequential addition of propylene and ethylene oxides. A range of materials is thereby available with a variety of properties determined by the ratio of the propylene and ethylene oxides in the polymer. They can be synthesised in various forms: liquid, paste or flaked solid. One of the forms of Pluronic polyol suggested for use as a wound dressing is F127 which forms a gel in water. The gels that form are temperature dependent in as much as F127 is more soluble in cold water than in hot due to hydrogen bond formation. If applied to a wound surface in cold solution and allowed to warm to body temperature, a gel is formed totally enclosing the wound and sealing it off from the environment. Antibacterial agents can be incorporated into the gel and so provide opportunities for designing a dressing suitable for clinical needs.

DESIGN OF AN ARTIFICIAL SKIN

Historically, dressings for wounds have been chosen empirically from available materials, but the growth of polymer chemistry has reached the stage where design of materials has become possible. Not only must suitable dressings protect the wound, control water vapour loss and assist or permit natural healing, but also must inhibit wound contracture. To do all this, the material must have the appropriate material properties, be compatible with tissues and, if possible, have antibacterial qualities. It is unlikely, therefore, that all the criteria will be met at one and the same time and the need for development suggests that a series of design stages would be appropriate.

The first stage in the design of a material for wound dressing is to satisfy basic needs relating to material properties. These may be summarised as: tensile strength both wet and dry, bending rigidity, tear strength, surface energy, moisture flux rate and blood compatibility. Once these criteria have been satisfied the next stage will involve animal and clinical tests to establish: graft/wound properties, wetting and adhesion, wound bed properties, viability of tissue, interface control, infection and fluid loss.

If under trial conditions a material is found which appears satisfactory, then a second series of clinical assessments and measurements have to be made: biodegradability rate, antigenicity, concentration changes in toxic metabolites, and migration of membrane particles or components. Finally, measurements have to be taken to determine the migration rate of non-inflammatory cells,

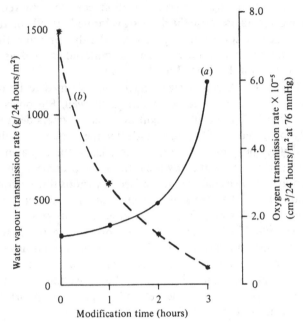

Fig. 56. Plot of (a) water vapour transmission rate against modification time of membrane by ethylene oxide, and (b) oxygen permeability rate against membrane modification time. (By permission of Dr J. M. Courtney.)

including new epithelium, rate of synthesis of dermal tissue, control of contracture and scar formation and, eventually, metabolic disposal of the graft.

A number of suitable membranes have already been designed which meet many of the above criteria. One of these is a copolymer film based on acrylonitrile (AN) and dimethylaminoethylmethacrylate (DMAEMA). It can be cast onto a polyester film substrate from a 20% w/v solution of the copolymer (85% AN, 15% DMAEMA) in dimethylformamide (DMF) (Park *et al.*, 1977).

Water vapour transmission through AN–DMAEMA copolymer membranes may be modified with ethylene oxide. The modification process takes place in a pressure chamber containing 100% ethylene oxide gas at a slight negative pressure (Courtney & Hood, 1976). Temperature in the chamber was maintained at $30 \pm 0.5\,^{\circ}\mathrm{C}$. Water vapour transmission was measured by placing a desiccant (calcium chloride) in a dish over which was stretched the membrane. The upper surface of the membrane was maintained in a moist state until the whole rig was placed in an incubator at $37\,^{\circ}\mathrm{C}$ and c. 70% relative humidity. The weight gain of the desiccant from water vapour derived from the moist atmosphere gave values for rates of vapour transmission. A similar rig was used to measure permeability to oxygen and carbon dioxide.

If water vapour transmission is plotted against modification time (Fig. 56a) an increase in transmission is noted with increasing exposure to ethylene oxide. The range of transmission rates can readily be selected between the limits 300 to 1200 g/m^2 per 24 hours. Normal skin has values between 150 and 450 g/m^2 per 24 hours. The plot of oxygen permeability against modification time shows a rapid decrease from 7.5×10^5 to 0.5×10^5 cm^3/m^2 per 24 hours at standard temperature and pressure (Fig. 56b). Carbon dioxide permeability falls within the range 1×10^5 to 16×10^5 cm^3/m^2 per 24 hours. The decrease in oxygen permeability was suggested to be dependent upon the degree of cross-linking in the membrane which causes a stiffer and more occlusive dressing despite the increase in hydrophilicity.

Bacteriological control may be affected by incorporating an antibacterial agent in the membrane itself. Tests with neomycin, tetracycline, sulphadiazine and silver sulphadiazine showed that the last was the most useful agent and, therefore, most suitable for incorporation in a wound dressing (Park & Mollison, 1978). Tests *in vitro* showed that 10% w/w of this agent could afford control of bacterial growth.

A further modification of the membranes can be achieved by adding different types of plasticisers. Park *et al.* (1978) tried two types of plasticisers with the AN–DMAEMA films: one a liquid, glycerol, the other a solid, Pluronic F68, a non-ionic surfactant based on a copolymer of ethylene oxide and propylene oxide. Varying the amounts of plasticiser between the limits 2% and 10% v/w or w/w, no change in the water vapour transmission data was found when Pluronic F68 was used but a significant increase was achieved with glycerol. In these experiments an additional method was used to measure water transmission. Water instead of the desiccant was placed in the dish below the test sample and the rig then put in the incubator at 37 °C but at a humidity of only *c*. 40%. The rate of loss of water through the membrane determined gravimetrically gave the transmission rate. Inevitably, higher transmission rates were achieved by this method but, as the authors suggest, since a true wound is neither completely dry nor a true liquid surface, values between the extremes measured by the two methods probably represent more accurately the clinical condition.

One of the disadvantages of the membrane system developed lies in its mechanical properties. In both dry and wet states the membrane is considerably less extensible than heavily plasticised PVC (Clingfilm, 'Stretch and Seal', etc.) and Opsite Mk. 2. In the wet state the peak loads and rupture loads are considerably less than in the dry state, but this is probably not so important as the strain parameter since pig skin is used successfully as a dressing and this has a rupture stress of only *c*. 8 kg/cm^2.

As yet, no satisfactory membrane that meets even the first stage requirements has been designed. The work of Yannas & Burke (1980) promises to provide a significant advance on the present state of the art. Basing their study on collagen as the framework material, they have used glycosaminoglycan

(GAG) as the second macromolecular component. Collagen alone, even though it can be reconstituted with a wide range of properties by adjusting the cross-link density, is still a relatively stiff and brittle material. By co-precipitating collagen with one of the several glycosaminoglycans (chondroitin-6-sulphate, chondroitin-4-sulphate, dermatan sulphate, heparin or heparin sulphate) membranes are obtained which are significantly more resistant to collagenase degeneration, have a higher modulus of elasticity and a higher energy of fracture. By microscopy, collagen–glycosaminoglycan membrane also has a more open pore structure than simple collagen membranes.

A further advantage was found in the system of precipitation. Collagen precipitated with chondroitin sulphate at pH 3.2 does not aggregate platelets and appears to prolong clotting time, whereas if dialysed to neutral pH it initiates aggregation. By adjusting the pH it should be possible to control blood-clotting time at the membrane/wound bed layer and so control the generation of the boundary layer.

At the present stage of development, manufactured membranes, when appropriately designed, can satisfy the wide range of criteria listed above, but no single unique solution has yet been devised. Already membranes that strongly retard wound contracture are available. When applied to wound sites they are replaced by newly synthesised, stable connective tissue. Already several rules relating the molecular structure and morphology of these membranes to cellular response of adjacent tissue have been established and a variety of interim solutions give satisfactory clinical results. We still await, however, the design of a truly compatible artificial skin which provides the unique solution being sought.

REFERENCES

Adachi, K., Takayasu, S., Takashima, I., Kano, M. & Kondo, S. (1970). Human hair follicles: metabolism and growth mechanisms. *Journal of the Society of Cosmetic Chemists*, **21**, 901–24.

Adams, T., Funkhouser, G. E. & Kendall, W. W. (1963). Measurement of evaporative water loss by a thermal conductivity cell. *Journal of Applied Physiology*, **18**, 1291–3.

Alexander, H. & Cook, T. H. (1977). Accounting for natural tension in the mechanical testing of human skin. *Journal of Investigative Dermatology*, **69**, 310–14.

Alexander, H. & Miller, D. L. (1979). Determining skin thickness with pulsed ultrasound. *Journal of Investigative Dermatology*, **72**, 17–19.

Alexander, J. W., Wheeler, L. M., Rooney, R. C., McDonald, J. J. & MacMillan, B. G. (1973). Clinical evaluation of Epigard, a new synthetic substitute for homograft and heterograft skin. *Journal of Trauma*, **13**, 374–82.

Al-Haboubi, A. M. (1977). Effect of different aqueous solutions on the physical and mechanical properties of connective tissues. MSc thesis, University of Strathclyde, Glasgow.

Alhopuro, S., Rintala, A., Salo, H. & Ritsila, V. (1976). Tissue adhesive vs. sutures in closure of incision wounds. A comparative study in human skin. *Annales Chirurgiae et Gynaecologiae (Helsinki)*, **65**, 308–12.

Allen, J. A., Armstrong, J. E. & Roddie, I. C. (1973). The regional distribution of emotional sweating in man. *Journal of Physiology*, **255**, 749–59.

Apolinar, E. & Rowe, W. F. (1980). Examination of human finger nail ridges by means of polarised light. *Journal of the Forensic Sciences*, **25**, 154–61.

Arem, A. J. & Kischer, C. W. (1980). Analysis of striae. *Plastic and Reconstructive Surgery*, **65**, 22–9.

Arem, A. J. & Madden, J. W. (1976). Effect of stress on healing wounds: I. Intermittent noncyclical tension. *Journal of Surgical Research*, **20**, 93–102.

ASTM Standards (1972). Standard methods of test for water vapour transmission of materials in sheet form. *Annual Book of ASTM Standard, part 14*. USA.

Astore, I. P. L., Pecoraro, V. & Pecoraro, E. G. (1979). The normal trichogram of pubic hair. *British Journal of Dermatology*, **101**, 441–4.

Baden, H. P. (1970). The physical properties of nail. *Journal of Investigative Dermatology*, **55**, 115–22.

Bailey, A. (1968). *International Review of Connective Tissue Research*, vol. 4, ed. D. A. Hall & D. S. Jackson pp. 233–79. Academic Press, New York & London.

Baird, H. W. (1964). Kindred showing congenital absence of the dermal ridges (fingerprints) and associated anomalies. *Journal of Pediatrics*, **64**, 621–31.

Baker, H. & Kligman, A. M. (1967). Measurement of transepidermal water loss by electrical hygrometry. *Archives of Dermatology*, **96**, 441–52.

Bakermann, S. (1964). Distribution of the α- and β-components in human skin collagen with age. *Biochimica et Biophysica Acta*, **90**, 621–3.

Barman, J. M., Astore, I. & Pecoraro, V. (1965). The normal trichogram of the adult. *Journal of Investigative Dermatology*, **44**, 233–6.

Barman, J. M., Pecoraro, V. & Astore, I. (1964). Method, technic and computations in the study of the trophic state of the human scalp hair. *Journal of Investigative Dermatology*, **42**, 421–5.

Barman, J. M., Pecoraro, V., Astore, I. & Ferrer, J. (1967). The first stage in the natural history of the human scalp. *Journal of Investigative Dermatology*, **48**, 138–42.

Barnett, A. (1938). The phase angle of normal human skin. *Journal of Physiology*, **93**, 349–66.

Barsky, S. H., Rosen, S., Geer, D. & Noe, J. M. (1980). The nature and evolution of port-wine stains: a computer-assisted study. *Journal of Investigative Dermatology*, **74**, 154–7.

Bearn, E. & Barsby, J. R. (1977). *Improvements in Dressings*, no. 54. British Patent Specification No. 1490065.

Bennett, J. P., Mathews, R. & Faulk, W. P. (1980). Treatment of chronic ulceration of the legs with human amnion. *Lancet*, **ii**, 1153–6.

Bergstresser, P. R. & Chapman, S. L. (1980). Maturation of normal human epidermis without an ordered structure. *British Journal of Dermatology*, **102**, 641–8.

Bergstresser, P. R., Pariser, R. J. & Taylor, J. R. (1978). Counting and sizing of epidermal cells in normal human skin. *Journal of Investigative Dermatology*, **70**, 280–4.

Bergstresser, P. R. & Taylor, J. R. (1977). Epidermal 'turnover time': a new examination. *British Journal of Dermatology*, **96**, 503–6.

Bernstein, B., Kearsley, E. A. & Zapas, L. S. (1963). A study of stress relaxation with finite strain. *Transactions of the Society for Rheology*, **7**, 391–410.

Bettley, F. R. (1961). Influence of soap on the permeability of the epidermis. *British Journal of Dermatology*, **73**, 448–54.

Bettley, F. R. (1963). Irritant effect of soap in relation to epidermal permeability. *British Journal of Dermatology*, **75**, 113–16.

Bettley, F. R. (1965). The influence of detergents on epidermal permeability. *British Journal of Dermatology*, **77**, 98–100.

Bettley, F. R. & Donaghue, E. (1960). Effect of soap on the diffusion of water through isolated human epidermis. *Nature (London)*, **185**, 17–20.

Bettley, F. R. & Grice, K. A. (1965). A method of measuring the transepidermal water loss and a means of inactivating sweat glands. *British Journal of Dermatology*, **77**, 627–38.

Bettley, R. & Grice, K. A. (1967). The influence of ambient humidity on the transepidermal water loss. *British Journal of Dermatology*, **78**, 475–81.

Bhangoo, K. S. & Church, J. C. (1976). Elastogenesis in healing wounds in bats. *Plastic and Reconstructive Surgery*, **57**, 468–79.

Bishop, J. C. (1979). DNA repair synthesis in human skin exposed to ultraviolet radiation used in PUVA (psoralen and UV-A) therapy for psoriasis. *British Journal of Dermatology*, **101**, 399–405.

Black, C. M., Gathercole, L. J., Bailey, A. J. & Beighton, P. (1980). The Ehlers–Danlos syndrome: an analysis of the structure of the collagen fibres of the skin. *British Journal of Dermatology*, **102**, 85–96.

Black, M. M. (1969). A modified radiographic method for measuring skin thickness. *British Journal of Dermatology*, **81**, 661–6.

Black, M. M. & Shuster, S. (1971). The physical properties of skin in cutis laxa. *British Journal of Dermatology*, **85**, 598–9.

Blank, I. H. (1952). Factors which influence the water content of the stratum corneum. *Journal of Investigative Dermatology*, **18**, 433–40.

Blank, I. H. (1953). Further observations on factors which influence the water content of the stratum corneum. *Journal of Investigative Dermatology*, **21**, 259–71.

Blank, I. H. & Gould, E. (1961). Penetration of anionic surfactants into skin. II. Study of mechanisms which impede the penetration of synthetic anionic surfactants into skin. *Journal of Investigative Dermatology*, **37**, 311–15.

Blank, I. H., Gould, E. & Theobald, A. (1964). Penetration of cationic surfactants into skin. *Journal of Investigative Dermatology*, **42**, 363–6.

Blank, I. H. & Shappirio, E. B. (1955). Water content of the stratum corneum. III. Effect of previous contact with aqueous solutions of soaps and detergents. *Journal of Investigative Dermatology*, **25**, 391–401.

Bodey, A. S. (1978). Structural changes in the skin occurring in Antarctica. *Clinical and Experimental Dermatology*, **3**, 417–23.

Boldingh, W. H. & Laurence, E. B. (1968). Extraction, purification and preliminary characterisation of the epidermal chalone. *European Journal of Biochemistry*, **5**, 191–8.

Bouissou, H., Pieraggi, M.Th., Julian, M. & Blazy, D. L. (1977). Histologie cutanée et arteriosclerose. *Revue de Medecine*, **28**, 1818–27.

Breathnach, A. S. (1964). Electron microscopy of melanocytes and melanosomes in freckled human epidermis. *Journal of Investigative Dermatology*, **42**, 339–94.

Breathnach, A. S. (1977). Electron microscopy of cutaneous nerves and receptors. *Journal of Investigative Dermatology*, **69**, 8–26.

Breathnach, A. S., Silvers, W. K., Smith, J. & Heyner, S. (1968). Langerhans cells in normal skin experimentally deprived of its neural crest component. *Journal of Investigative Dermatology*, **50**, 147–60.

Brenglemann, G. L., McKeag, M. & Rowell, L. B. (1975). Use of dew-point detection for quantitative measurement of sweating rate. *Journal of Applied Physiology*, **39**, 498–500.

Brereton, W. D. S. (1974). Skin extensibility and its orientational variation *in vivo*: a suction technique. MSc thesis, University of Strathclyde, Glasgow.

Briggaman, R. & Wheeler, C. E. (1975). The epidermal–dermal junction. *Journal of Investigative Dermatology*, **65**, 71–84.

Brodsky, B., Eikenberry, E. F. & Cassidy, K. (1980). An unusual collagen periodicity in skin. *Biochimica et Biophysica Acta*, **621**, 162–86.

Bronstad, G. O., Elgjo, K. & Oye, I. (1971). Adrenaline increases cyclic $3',5'$-AMP formation in hamster epidermis. *Nature (London)*, **233**, 78–9.

Brown, A. G., Fyffe, R. E. W. & Noble, R. (1980). Projections from Pacinian corpuscles and rapidly adapting mechanoreceptors of glabrous skin to the cat's spinal cord. *Journal of Physiology*, **307**, 385–400.

Brown, I. A. (1971). Structural aspects of the biomechanical properties of human skin. PhD thesis, University of Strathclyde, Glasgow.

Brown, I. A. (1973). A scanning electron microscope study of the effects of increased tension on human skin. *British Journal of Dermatology*, **89**, 383–93.

Brunius, U. & Ahren, C. (1969). Healing of skin incisions during reduced tension. *Acta Chirurgica Scandinavica*, **135**, 383–90.

Buettner, R. (1936). The influence of blood circulation on the transport of heat in the skin. *Strahlentherapie*, **55**, 333. (Quoted by Cohen, 1977.)

Bull, J. P. (1963). Burns. *Postgraduate Medical Journal*, **39**, 717–23.

Bullough, W. S. & Laurence, E. B. (1960). The control of epidermal mitotic activity in the mouse. *Proceedings of the Royal Society of London Series B*, **151**, 536.

Bullough, W. S. & Laurence, E. B. (1964). Mitotic control by internal secretion. The role of the chalone–adrenaline complex. *Experimental Cell Research*, **33**, 176–94.

Bunting, C. H. & Eades, C. (1926). Effects of mechanical tension on the polarity of growing fibroblasts. *Journal of Experimental Medicine*, **44**, 147–49.

Burbank, D. P. & Webster, J. G. (1978). Reducing skin potential motion artefact by skin abrasion. *Medical and Biological Engineering and Computing*, **16**, 31–8.

Burge, K. M. & Winklemann, R. K. (1970). Mercury pigmentation: an electron microscope study. *Archives of Dermatology*, **102**, 51–61.

Burlin, T. E., Hutton, W. C. & Ranu, H. S. (1977). A method of *in vivo* measurement of the elastic properties of skin in radiotherapy patients. *Journal of Investigative Dermatology*, **69**, 321–3.

Burton, J. L. (1972). Factors affecting the rate of sebum excretion in man. *Journal of the Society of Cosmetic Chemists*, **23**, 241–58.

Burton, J. L. & Shuster, S. (1973). A rapid increase in skin extensibility due to prednisolone. *British Journal of Dermatology*, **89**, 491–5.

Callender, R. M. (1974). The optical texture of human skin. *Medical and Biological Illustration*, **24**, 171–3.

Campbell, S. D., Kraning, K. K., Schibli, E. G. & Momii, S. T. (1977). Hydration characteristics and electrical resistivity of stratum corneum using a four point microelectrode method. *Journal of Investigative Dermatology*, **69**, 290–5.

Carlsen, R. A. & Asboe-Hansen, G. (1971). Changes in skin histamine induced by X-rays of differing quality. *Journal of Investigative Dermatology*, **56**, 69–71.

Carpendale, M. T. (1964). Reduction of surgical wound infection by tape closure. *Surgical Forum*, **15**, 58–60.

Carton, R. W., Dainauskas, J. & Clarke, J. W. (1962). Elastic properties of single elastic fibers. *Journal of Applied Physiology*, **17**, 547–51.

Champion, R. H. (1970). Sweat glands. In *An Introduction to the Biology of Skin*, ed. R. H. Champion, T. Gilman, A. J. Rook & R. T. Sims, chapt. 12. Blackwell Scientific, Oxford.

Champion, R. H., Gilman, T., Rook, A. J. & Sims, R. T. (1970). *An Introduction to the Biology of the Skin*. Blackwell Scientific, Oxford.

Chapman, L. F. (1977). Mechanisms of the flare reaction in human skin. *Journal of Investigative Dermatology*, **69**, 88–97.

Chardack, W. M., Day, C. W., Fazekas, G. & Muisley, N. (1961). Synthetic skin: an experimental study. *Journal of Trauma*, **1**, 54–8.

Chatterjee, D. S. (1978). A controlled comparative study of the use of porcine xenograft in the treatment of partial thickness skin loss in an occupational health centre. *Current Medical Research and Opinion*, **5**, 726–33.

Chernovsky, M. E. & Knox, J. M. (1964). Atrophic striae after occlusive corticosteroid therapy. *Archiv Dermatologica*, **90**, 15–19.

Cherry, R. H. (1965). Thermal conductivity gas analysis in hygrometric applications. In *Humidity and Moisture*, vol. 1, ed. A. Wexter & R. E. Ruskin. Reinhold, New York.

Chieffi, M. (1950). An investigation of the effects of parenteral and topical administration of steroids on the elastic properties of senile skin. *Journal of Gerontology*, **5**, 17–22.

Chopra, D. P. & Flaxman, B. A. (1974). Comparative proliferation kinetics of cells from normal human epidermis and benign epidermal hyperplasia (psoriasis) *in vitro*. *Cell and Tissue Kinetics*, **7**, 69–76.

Christensen, M. S., Hargens, C. W., Nacht, S. & Guns, E. H. (1977). Viscoelastic properties of intact human skin: instrumentation, hydration effects, and the contribution of the stratum corneum. *Journal of Investigative Dermatology*, **69**, 282–6.

Clar, E. J., Her, C. P. & Sturelle, C. G. (1975). Skin impedance and moisturization. *Journal of the Society of Cosmetic Chemists*, **26**, 337–53.

Clark, R. P., Mullan, B. J. & Pugh, L. G. C. E. (1977). Skin temperature during running: a study using infra-red colour thermography. *Journal of Physiology*, **267**, 53–62.

Clark, R. P. & Stothers, J. K. (1980). Neonatal skin temperature distribution using infra-red colour thermography. *Journal of Physiology*, **302**, 323–33.

Clough, J. V. & Alexander-Williams, J. (1975). Surgical and economic advantages of polyglycolic acid suture material in skin closure. *Lancet*, **i**, 194.

Cochrane, R. G. (1968). The low temperature storage of skin: a preliminary report. *British Journal of Plastic Surgery*, **21**, 118–25.

Cohen, I. (1979). Stress and wound healing. *Acta Anatomica*, **103**, 134–41.

Cohen, M. L. (1977). Measurement of the thermal properties of human skin: a review. *Journal of Investigative Dermatology*, **69**, 333–8.

Cohen, R. E., Hooley, C. J. & McCrum, N. G. (1976). Visco-elastic creep of collagenous tissues. *Journal of Biomechanics*, **9**, 175–84.

Cohen, S. (1966). An investigation and fractional assessment of the evaporative water loss through normal skin and burn eschars using a microhygrometer. *Plastic and Reconstructive Surgery*, **37**, 475–86.

Cole, K. S. (1940). Permeability and impermeability of cell membranes for ions. *Cold Spring Harbor Symposia on Quantitative Biology*, **8**, 110–22.

Cole, K. S. & Cole, R. H. (1941). Dispersion and absorption in dielectrics. I. Alternating current characteristics. *Journal of Chemical Physics*, **9**, 341–51.

Colgan, D. M. (1970). Effects of instructions on the skin resistance response. *Journal of Experimental Psychology*, **86**, 108–12.

Comninou, M. & Yannas, I. V. (1976). Dependence of stress–strain non-linearity of connective tissues on the geometry of collagen fibres. *Journal of Biomechanics*, **9**, 427–33.

Cook, T., Alexander, H. & Cohen, M. (1977). Experimental method for determining the 2-dimensional mechanical properties of living human skin. *Medical and Biological Engineering and Computing*, **15**, 381–90.

Cooper, J. H. (1969). Histochemical observations on the elastic sheath: elastofibril system of the dermis. *Journal of Investigative Dermatology*, **52**, 169–76.

Courtney, J. M., Gilchrist, T., Park, G. B. & Lindsay, R. M. (1975). Blood compatibility and acrylonitrile copolymer structure. *Proceedings of the European Society for Artificial Organs*, **2**, 156–60.

Courtney, J. M. & Hood, R. G. (1976). The design of a gas reactor for modifying and controlling the properties of polymers. *Engineering Designer*, **2**, 13–17.

Cowdry, C. K. & Cowdry, E. V. (1950). Aging of elastic tissue in human skin. *Journal of Gerontology*, **5**, 203–10.

Cowles, M. P. (1973). The latency of the skin resistance response and reaction time. *Psychophysiology*, **10**, 177–83.

Cox, H. T. (1941). The cleavage lines of the skin. *British Journal of Surgery*, **19**, 234–40.

Craik, J. E. & McNeil, I. R. R. (1965). Histological studies of stressed skin. In *Biomechanics and Related Bio-engineering Topics*, ed. R. M. Kenedi, pp. 159–64. Pergamon Press, Oxford.

Cruikshank, C. N. D. & Hershey, F. B. (1960). The effects of heat on the metabolism of guinea pig ear skin. *Annals of Surgery*, **151**, 419–30.

Cummins, H. (1964). Dermatoglyphics: a brief review. In *The Epidermis*, ed. W. Montagna & W. Lobitz, pp. 375–86. Academic Press, New York & London.

Cummins, H. & Midlo, C. (1943). *Fingerprints, Palms and Soles, An Introduction to Dermatoglyphics*. Blakiston, Philadelphia.

Daly, C. H. (1966). The biomechanical characteristics of human skin. PhD thesis, University of Strathclyde, Glasgow.

Daly, C. H. & Odland, G. F. (1979). Age-related changes in the mechanical properties of human skin. *Journal of Investigative Dermatology*, **73**, 84–7.

Darian-Smith, I. & Johnson, K. O. (1977). Thermal sensibility and thermoreceptors. *Journal of Investigative Dermatology*, **69**, 146–53.

Davis, J. W. (1910). Skin transplantation with a review of 550 cases at The Johns Hopkins Hospital. *Johns Hopkins Medical Journal*, **15**, 307–10.

Dawber, R. P. R. (1977). Weathering of hair in monilethrix and pili torti. *Clinical and Experimental Dermatology*, **2**, 271–7.

Dawber, R. P. R., Marks, R. & Swift, J. A. (1972). Scanning electron microscopy of the stratum corneum. *British Journal of Dermatology*, **86**, 272–81.

Dehoff, P. H. (1978). On the nonlinear viscoelastic behaviour of soft biological tissues. *Journal of Biomechanics*, **11**, 35–40.

Delforno, C., Holt, P. J. A. & Marks, R. (1978). Corticosteroid effect on epidermal cell size. *British Journal of Dermatology*, **98**, 619–23.

Demiray, H. (1972). A note on the elasticity of soft biological tissues. *Journal of Biomechanics*, **5**, 309–11.

Dempsey, M. (1979). Scanning electron microscope studies of some normal and diseased conditions of mammalian skin as revealed in leather manufacture. *Journal of Pathology*, **128**, 151–7.

Derksen, W. L., Murtha, T. D. & Monahan, T. I. (1957). Thermal conductivity and diathermancy of human skin for sources of intense thermal radiation employed in flash burn studies. *Journal of Applied Physiology*, **11**, 205–10.

Diamant, J., Keller, A., Baer, E., Litt, M. & Arridge, R. G. C. (1972). Collagen: ultrastructure and its relation to mechanical properties as a function of ageing. *Proceedings of the Royal Society of London, Series B*, **180**, 293–315.

Dick, J. C. (1951). The tension and resistance to stretching of human skin and other membranes, with results from a series of normal and oedematous cases. *Journal of Physiology*, **112**, 102–13.

Dioguardi, D. & Musajo-Somma, A. (1977). Threaded tapes for sutureless closure of skin wounds. *British Journal of Plastic Surgery*, **30**, 202–5.

Donovan, K. M. (1977). Antarctic environment and nail growth. *British Journal of Dermatology*, **96**, 507–10.

Douglas, B. (1952). Homografts of foetal membranes as a covering for large wounds – especially those from burns. An experimental and clinical study. *Tennessee State Medical Journal*, **45**, 230–7.

Dupuytren, G. (1836). *Theoretisch-praktische Vorlesungen über der Verletzungen durch Kriegswaffen.* (Originally published as *Traité théorique et pratique des blessures par armes de guerre.*) Veit, Berlin.

Dykes, P. J. & Marks, R. (1979). An appraisal of the methods used in the assessment of atrophy from topical corticosteroids. *British Journal of Dermatology*, **101**, 599–609.

Dykes, P. J., Marks, R. & Blakemore, C. B. (1978). Assessment of atrophogenicity of popularly prescribed topical corticosteroids. *British Journal of Clinical Practice*, **32**, 345–8.

Ebling, F. J. (1972). Does the pituitary gland affect the secretion of sebum? *Journal of the Society of Cosmetic Chemists*, **23**, 393–404.

Ebling, F. J. (1974). Sex hormones and skin. *Journal of the Society of Cosmetic Chemists*, **25**, 381–95.

Edelberg, R. (1968). Biopotentials from the skin surface: the hydration effect. *Annals of the New York Academy of Sciences*, **148**, 252–62.

Edelberg, R. (1971). Electrical properties of skin. In *Biophysical Properties of the Skin*, ed. H. R. Elden pp. 513–50. Wiley-Interscience, New York.

Edelberg, R. (1977). Relation of electrical properties of skin to structure and physiologic state. *Journal of Investigative Dermatology*, **69**, 324–7.

Edwards, R. P. (1975). Longspacing collagen in skin biopsies from patients in lepromatous leprosy. *British Journal of Dermatology*, **93**, 175–82.

Elfbaum, S. G. & Wolfram, M. A. (1970). Effect of dimethyl sulfoxide and other reagents upon mechanical properties of stratum corneum strips. *Journal of the Society of Cosmetic Chemists*, **21**, 129–40.

Eller, J. J. & Eller, W. D. (1949). Oestrogenic ointments. Cutaneous effects of topical applications of natural oestrogens with report of three hundred and twenty-one biopsies. *Archiv für Dermatologie und Syphilis*, **59**, 449–55.

Elliot, R. A. & Hoehn, J. G. (1973). Use of commercial porcine skin for wound dressings. *Plastic and Reconstructive Surgery*, **52**, 401–5.

Engesaeter, L. B. & Skar, A. G. (1978). Effects of oxytetracycline on the mechanical properties of bone and skin in young rats. *Acta Orthopaedica Scandinavica*, **49**, 529–34.

Engesaeter, L. B. & Skar, A. G. (1979). Effects of cloxicillin, doxycycline, fusidic acid and lincomycin on the mechanical properties of bone and skin in young rats. *Acta Orthopaedica Scandinavica*, **50**, 245–9.

Enna, C. D. & Dyer, R. F. (1979). The histomorphology of the elastic tissue system in the skin of the human hand. *Hand*, **11**, 144–50.

Ersek, R. A. & Mayer, M. H. (1980). Treatment of avulsion injuries with porcine skin xenografts. *Surgery, Gynecology and Obstetrics*, **151**, 33–5.

Evans, J. H. & Barbenel, J. C. (1975). Structural and mechanical properties of tendon related to function. *Equine Vetinarian*, **7**, 1–8.

Everett, M. A., Yeargers, E., Sayre, R. M. & Olson, R. J. (1966). Penetration of epidermis by ultraviolet rays. *Photochemistry and Photobiology*, **5**, 533–42.

Facq, J., Kirk, D. I. & Rebell, G. (1964). A simple replica technique for the observation of human skin. *Journal of the Society of Cosmetic Chemists*, **15**, 87–9.

Fang, T. C. (1950). The inheritance of the a–b ridge-count on the human palm with a note on its relation to mongolism. PhD thesis, University of London.

Ferguson, J. & Barbenel, J. C. (1981). Skin surface patterns and the directional mechanical properties of the dermis. In *Bioengineering and the Skin*, ed. R. Marks & P. A. Payne pp. 83–92. MTP, Lancaster.

Findlay, G. H. (1970). Blue skin. *British Journal of Dermatology*, **83**, 127–34.

Finean, J. B. & Millington, P. F. (1955). Low angle X-ray diffraction studies of the polymorphic forms of synthetic $\alpha:\alpha$ and $\alpha:\beta$ kephaline and $\alpha:\beta$ lecithins. *Transactions of the Faraday Society*, **51**, 1008–15.

Finlay, B. (1970). Dynamic mechanical testing of human skin *in vitro*. *Journal of Biomechanics*, **3**, 557–68.

Finlay, B. (1971). The torsional characteristics of human skin *in vivo*. *Biomedical Engineering*, **6**, 567–73.

Finlay, J. B. (1978). Thixotropy in living skin. *Journal of Biomechanics*, **11**, 333–42.

Fisher, L. B. (1968a). Determination of the normal rate and duration of mitosis in human epidermis. *British Journal of Dermatology*, **80**, 24–8.

Fisher, L. B. (1968b). The diurnal mitotic rhythm in the human epidermis. *British Journal of Dermatology*, **80**, 75–80.

Fisher, L. B. & Maibach, H. I. (1971). The effect of corticosteroids on human epidermal mitotic activity. *Archives of Dermatology*, **103**, 39–44.

Fisher, L. B. & Maibach, H. I. (1972). The effect of occlusive and semi-permeable dressings on the mitotic activity of normal and wounded human epidermis. *British Journal of Dermatology*, **86**, 593–600.

Flaxman, B. A. & Chopra, D. P. (1972). Cell cycle of normal and psoriatic epidermis *in vitro*. *Journal of Investigative Dermatology*, **59**, 102–5.

Foritz, H., Hepperde, W., Meyer, A., Pfeifer, M., Riedeberger, J. & Schumann, R. (1979). Producing a medicinal wound dressing. UK Patent, GB, 200288A.

Forrester, J. C., Zederfeldt, B. H., Hayes, T. L. & Hunt, T. K. (1970). Tape closed and sutured wounds: a comparison by tensiometry and scanning electron microscopy. *British Journal of Surgery*, **57**, 729–37.

Foster, K. G., Ginsburg, J. & Weiner, J. S. (1970). Role of circulating catecholamines in human eccrine sweat gland control. *Clinical Science*, **39**, 823–32.

Friedrich, L., Wuppermann, D. & Zimmermann, F. (1975). Influence of concomitant treatment with D-penicillamine and β-amino-aminoproprionitrile on the mechanical properties of rat connective tissue. *Archives of Dermatological Research*, **254**, 303–11.

Fry, P., Harkness, M. L. R. & Harkness, R. D. (1964). Mechanical properties of the collagenous framework of skin in rats of different ages. *American Journal of Physiology*, **206**, 1425–9.

Fry, P., Harkness, M. L. R., Harkness, R. D. & Nightingale, M. (1962). Mechanical properties of tissues of lathyritic animals. *Journal of Physiology*, **164**, 77–89.

Fujita, T. (1973). Surface structure of the skin. In *Basic Dermatology*, ed. K. Seijk & Y. Mishima. Asakura Shoten, Tokyo.

Fullmer, H. M. (1960). A comparative histochemical study of elastic, pre-elastic and oxytalin connective tissue fibres. *Journal of Histochemistry and Cytochemistry*, **8**, 290–5.

Fung, Y. C. (1968). Biomechanics: its scope, history and some problems of continuum mechanics in physiology. *Applied Mechanics Review*, **21**, 1–20.

Furchtgott, E. & Busemeyer, J. K. (1979). Heart rate and skin conductance during cognitive processes as a function of age. *Journal of Gerontology*, **34**, 183–90.

Galton, Sir F. (1892). *Finger Prints*. (Reprinted in 1965 by Da Capo Press, New York.)

Gasselt, H. R. M. & Vierhout, R. R. (1963). Registration of the insensible perspiration of small quantities of sweat. *Dermatologica (Basel)*, **127**, 255–9.

Gaul, L. E. & Underwood, G. B. (1952). Relation of dew point and barometric pressure to chapping of normal skin. *Journal of Investigative Dermatology*, **19**, 9–19.

Geschwandtner, W. R. (1973). Striae cutis atrophicae nach Lokalbehandlung mit Corticosteroiden. *Hautarzt*, **24**, 70–3.

Gibson, T., Kenedi, R. M. & Craik, J. E. (1965). The mobile microarchitecture of dermal collagen. *British Journal of Surgery*, **52**, 764–70.

Gillard, G. C., Reilly, H. C., Bell-Booth, P. G. & Flint. M. H. (1977). A comparison of the glycosaminoglycans of weight-bearing and non-weight-bearing human dermis. *Journal of Investigative Dermatology*, **69**, 257–61.

Gilman, T., Penn, J., Bronks, D. & Roux, M. (1953). Reactions of healing wounds and granulation tissue in man to auto-thiersch, auto-dermal and homo-dermal grafts. *British Journal of Plastic Surgery*, **6**, 153–223.

Goldman, J., Hornby, P. & Long, C. (1964). Effect of the laser on the skin. III. Transmission of the laser beams through fiber optics. *Journal of Investigative Dermatology*, **42**, 231–4.

Goldman, L. & Rockwell, R. J. (1971). *Lasers in Medicine*. Gordon & Breach Science Publications, New York.

Goldman, L., Vahl, J., Rockwell, R. J., Meyer, R., Franzen, M., Owens, P. & Hyatt, S. (1969). Replica microscopy and scanning electron microscopy of laser impacts on the skin. *Journal of Investigative Dermatology*, **52**, 18–24.

Goldsmith, L. A. & Baden, H. P. (1970). The mechanical properties of hair. I. The dynamic sonic modulus. *Journal of Investigative Dermatology*, **55**, 256–9.

Goldzieher, J. (1949). The direct effect of steroids on the senile human skin. *Journal of Gerontology*, **4**, 104–12.

Goldzieher, M. A. (1946). The effects of estrogens on the senile skin. *Journal of Gerontology*, **1**, 196–201.

Goodman, A. B. & Wolf, A. V. (1969). Insensible water loss from human skin as a function of ambient vapour concentration. *Journal of Applied Physiology*, **26**, 203–7.

Gopalan, C., Reddy, V. & Mohan, V. S. (1963). Some aspects of copper metabolism in protein calorie malnutrition. *Journal of Pediatrics*, **63**, 646–9.

Gosline, J. M. & French, C. J. (1979). Dynamic mechanical properties of elastin. *Biopolymers*, **18**, 2091–103.

Grahame, R. (1970). A method for measuring human skin elasticity *in vivo* with observations on the effects of age, sex and pregnancy. *Clinical Science*, **39**, 223–38.

Grahame, R. & Beighton, P. (1971). The physical properties of skin in cutis laxa. *British Journal of Dermatology*, **84**, 326–9.

Grandori, F. & Pedotti, I. (1980). Theoretical analysis of mechano-to-neural transduction in Pacinian corpuscle. *IEEE Transactions: Biomedical Engineering*, **27**, 559–65.

Grant, J. K. (1969). The metabolism of steroids in man. *British Journal of Dermatology*, **81**, *Supplement 2*, 18–22.

Green, A. E. & Zerna, W. (1954). *Theoretical Elasticity*. Oxford University Press.

Green, I., Stingl, G., Shevach, E. & Katz, S. I. (1980). Antigen presentation and allogenic stimulation by Langerhans cells. *Journal of Investigative Dermatology*, **75**, 44–5.

Grice, K., Sattar, H. & Baker, H. (1972). The effect of ambient humidity on transepidermal water loss. *Journal of Investigative Dermatology*, **58**, 343–6.

Grossman, M. I., Brazier, M. A. & Lechago, J. (eds.). (1981). *Cellular Basis of Chemical Messengers in the Digestive System*. Academic Press, New York & London.

Gunner, C. W., Hutton, W. C., Burlin, T. E. & Williams, E. W. (1979). The effects of some clinical treatments on the mechanics of skin. *Engineering in Medicine*, **8**, 58–62.

Hackett, M. E. J. (1975). Preparation, storage, and use of homograft. *British Journal of Hospital Medicine*, **13**, 272–84.

Hale, A. R. (1949). Breadth of epidermal ridges in the human fetus and its relation to the growth of the hand and foot. *Anatomical Record*, **105**, 763–76.

Hale, A. R. (1952). Morphogenesis of volar skin on the human fetus. *American Journal of Anatomy*, **91**, 147–81.

Hall, C. W., Liotta, D. & DeBakey, M. E. (1966). Artificial skin. *Transactions of the American Society for Artificial Internal Organs*, **12**, 340–5.

Halprin, K. M. (1972). Epidermal 'turnover time': a re-examination. *British Journal of Dermatology*, **86**, 14–19.

Hambrick, G. W. & Blank, H. (1954). Whole mounts for the study of skin and its appendages. *Journal of Investigative Dermatology*, **23**, 437–53.

Hammerlund, K., Nilsson, G. E., Oberg, P. A. & Sedin, G. (1977). Transepidermal water loss in newborn infants in relation to ambient humidity and site of measurement and estimation of total transepidermal water loss. *Acta Paediatrica Scandinavica*, **66**, 553–62.

Hanawalt, P. C., Cooper, P. K., Ganesan, A. K. & Smith, C. A. (1979). DNA repair in bacteria and mammalian cells. *Annual Reviews of Biochemistry*, **48**, 783–836.

Harburg, E., Gliebermann, L., Roeper, P., Schork, M. A. & Schiell, W. J. (1978a). Skin colour: ethnicity and blood pressure. I. Detroit blacks. *American Journal of Public Health*, **68**, 1177–83.

Harburg, E., Gliebermann, L., Ozgoren, F., Roeper, P. & Schork, M. A. (1978*b*). Skin colour: ethnicity and blood pressure. II. Detroit whites. *American Journal of Public Health*, **68**, 1184–88.

Harding, K. & Richardson, G. (1979). Silastic foam elastomer for treating open granulating wounds. *Nursing Times* 27 Sept.

Hardy, J. D., Wolff, H. G. & Godell, H. (1952). *Pain Sensations and Reactions*. New York, Hafner.

Harris, D. R., Filarski, S. S. & Hector, R. E. (1973). The effect of Silastic sheet dressings on the healing of split-skin graft donor sites. *Plastic and Reconstructive Surgery*, **52**, 189–92.

Harvey, W., Pope, F. M. & Grahame, R. (1975). Cutaneous extensibility in pesudo-xanthoma elasticum. *British Journal of Dermatology*, **97**, 679–82.

Hashimoto, K. (1974). New methods for surface ultrastructure: comparative studies of scanning electron microscopy, transmission electron microscopy and replica methods. *International Journal of Dermatology*, **13**, 357–81.

Hashimoto, K. & Kanzaki, T. (1975). Surface ultrastructure of human skin. *Acta Dermatovener (Stockholm)*, **55**, 413–30.

Hauser, I. (1938). Ueber spezifische Wirkungen des langwelliges ultravioletten Lichts auf die menschliche Haut. *Strahlentherapie*, **62**, 315–22.

Haut, R. C. & Little, R. W. (1972). A constitutive equation for collagen fibres. *Journal of Biomechanics*, **5**, 423–30.

Hell, E. & Lawrence, J. C. (1979). The initiation of epidermal wound healing in cuts and burns. *British Journal of Experimental Pathology*, **60**, 171–9.

Henderson, G. H., Karg, G. M. & O'Neill, J. J. (1978). Fractography of human hair. *Journal of the Society of Cosmetic Chemists*, **29**, 449–67.

Henriques, F. C. & Moritz, A. R. (1947). Studies of thermal injury. I. The conduction of heat to and through skin and the temperatures attained therein. A theoretical and an experimental investigation. *American Journal of Pathology*, **23**, 531–49.

Henschke, U. & Schulze, R. (1939). Untersuchungen zum Problem der ultraviolett-Dosimetrie. III. Ueber Pigmentierung durch langwelliges Ultraviolett. *Strahlentherapie*, **64**, 14–42.

Henshaw, P. S., Snider, R. S. & Riley, E. F. (1949). Aberrant tissue developments in rats exposed to beta rays. *Radiology*, **52**, 401–7.

Herrick, E. H. (1945). Tensile strength of tissues as influenced by male sex hormones. *Anatomical Record*, **73**, 145–9.

Herrick, E. H. & Brown, K. (1952). Lowered tensile strength and collagen levels in tissues following discontinuation of male sex hormone. *Poultry Science*, **31**, 191–3.

Heyden, B. (1969). Uber die Innervation der behaarten Haut des Menschen. *Acta Anatomica*, **74**, 20–9.

Hildebrandt, J., Fukaya, H. & Martin, C. J. (1969). Completing the length–tension curve of tissue. *Journal of Biomechanics*, **2**, 463–7.

Holbrook, K. A. & Odland, G. F. (1974). Regional differences in the thickness (cell layers) of the human stratum corneum: an ultrastructure analysis. *Journal of Investigative Dermatology*, **62**, 415–22.

Hooper, C. E. (1961). Use of colchicine for the measurement of mitotic rate in the intestinal epithelium. *American Journal of Anatomy*, **108**, 231–44.

Hoppe, U. (1979). Topologie der Hautoberflache. *Journal of the Society of Cosmetic Chemists*, **30**, 213–38.

Horch, K. W., Tuckett, R. P. & Burgess, P. R. (1977). A key to the classification of cutaneous mechano-receptors. *Journal of Investigative Dermatology*, **69**, 75–82.

Houck, J. C. (1968). A personal overview of inflammation. *Biochemical Pharmacology*, *Supplement*, 1–3.

Hrdy, D. (1973). Quantitative hair form variation in seven populations. *American Journal of Physical Anthropology*, **39**, 7–11.

Hunter, R., Pinkus, H. & Steele, C. H. (1956). Examination of the epidermis by the strip method. II. The number of keratin cells in the human epidermis. *Journal of Investigative Dermatology*, **27**, 31–4.

Hzuka, H., Kamigati, K., Nemoto, O., Aoyagi, T. & Miura, Y. (1980). Effects of hydrocortisone on the adrenaline–adenylate cyclase system of the skin. *British Journal of Dermatology*, **102**, 703–9.

Idson, B. (1973). Water and the skin. *Journal of the Society of Cosmetic Chemists*, **24**, 197–212.

Imokawa, G., Sumura, K. & Katsumi, M. (1975*a*). Study in skin roughness caused by surfactants. I. A new method for an *in vivo* evaluation of skin roughness. *Journal of the American Oil Chemists' society*, **52**, 479–83.

Imokawa, G., Sumura, K. & Katsumi, M. (1975*b*). Study in skin roughness caused by surfactants. II. Correlation between protein denaturation and skin roughness. *Journal of the American Oil Chemists' Society*, **52**, 479–83.

Ishida, T., Kashibuchi, N., Movita, J. & Yuasa, S. (1978). Measurements of skin roughness by computerised surface tracing and applications. In *Cosmetic Efficiency Substantiation*. IFSCC Congress, Australia.

Iverson, O. H. (1969). Chalones of the skin. In *Homeostatic Regulators*, ed. G. Wolstenholme & J. Knight pp. 29–55. Churchill, London.

Jackson, R. (1976). The lines of Blaschko: a review and reconsideration. *British Journal of Dermatology*, **95**, 349–60.

James, J. H. & Watson, A. C. H. (1975). The use of Opsite, a vapour permeable dressing, on skin graft donor sites. *British Journal of Plastic Surgery*, **28**, 107–10.

James, M. P., Black, M. M. & Sparkes, C. G. (1977). Measurement of dermal atrophy induced by topical steroids using a radiographic technique. *British Journal of Dermatology*, **96**, 303–66.

Jamison, C. E., Marangoni, R. D. & Glaser, A. A. (1968). Visco-elastic properties of soft tissue by discrete model characterisation. *Journal of Biomechanics*, **1**, 33–46.

Jansen, L. H. & Rottier, P. B. (1958). Some mechanical properties of human abdominal skin measured on excised strips. *Dermatologica*, **117**, 65–83.

Jarrett, A. (ed. (1973). *The Physiology and Pathophysiology of the Skin*, vol. 1, *Epidermis*. Academic Press, New York & London.

Jarrett, A. (ed. (1978). *The Physiology and Pathophysiology of the Skin*, vol. 5, *Sweat Glands, Skin Permeation, Lymphatics and the Nails*. Academic Press, New York & London.

Jimbow, W., Fitzpatrick, T. B., Szabo, G. & Hori, Y. (1975). Congenital circumscribed hypo-melanosis: a characterisation based on electron microscopic study of tuberous sclerosis, nevus depigmentosus, and piebaldism. *Journal of Investigative Dermatology*, **64**, 50–62.

Johansson, R. S. (1978). Tactile sensibility in the human hand: receptive field characteristics of mechano-receptive units in the glabrous skin area. *Journal of Physiology*, **281**, 101–23.

Johnson, C. & Shuster, S. (1969). The measurement of transepidermal water loss. *British Journal of Dermatology, Supplement*, **4**, 40–6.

Kaidbey, K. H. & Kligman, A. M. (1975). Further studies of photo-augmentation in humans: phototaxic reactions. *Journal of Investigative Dermatology*, **65**, 472–5.

Kaidbey, K. H. & Kligman, A. M. (1978). The acute effects of long-wave ultraviolet radiation on human skin. *Journal of Investigative Dermatology*, **72**, 253–6.

Kakivaya, S. R. & Hoeve, C. A. J. (1975). The glass point of elastin. *Proceedings of the National Academy of Sciences, USA*, **72**, 3505–7.

Keatinge, N. R. & Cannon, P. (1960). Freezing point of human skin. *Lancet*, **i**, 11–14.

Kenedi, R. M., Gibson, T. & Daly, C. H. (1965). Bio-engineering studies of the human skin. In *Biomechanics and Related Bio-engineering Topics*, ed. R. M. Kenedi, pp. 147–58. Pergamon Press, Oxford.

Kenedi, R. M., Gibson, T., Evans, J. H. & Barbenel, J. C. (1975). Tissue mechanics. *Physics in Medicine and Biology*, **20**, 699–717.

Kennedy, C., Molland, E., Henderson, W. J. & Whiteley, A. M. (1977). Mercury pigmentation from industrial exposure. *British Journal of Dermatology*, **96**, 367–74.

King, C. S., Nicholls, S. & Marks, R. (1981). Relationship of intracorneal cohesion to rates of desquamation in the scaling disorders. In *Bioengineering and the Skin*, ed. R. Marks & P. A. Payne, pp. 237–44. MTP, Lancaster.

Kirk, E. & Kvorning, S. A. (1949). Quantitative measurements of the elastic properties of the skin and subcutaneous tissue in young and old individuals. *Journal of Gerontology*, **4**, 273–84.

Kirk, J. E. & Chieffi, M. (1962). Variation with age in elasticity of skin and subcutaneous tissue in human individuals. *Journal of Gerontology*, **17**, 373–80.

Klein, E., Fine, S., Laor, Y., Litwin, M. S., Donaghue, J. & Simpson, L. (1964). Laser irradiation of the skin. *Journal of Investigative Dermatology*, **43**, 565–70.

Kligman, A. M. (1964). The biology of the stratum corneum. In *The Epidermis*, ed. W. Montagna & W. C. Lobitz, pp. 387–433. Academic Press, New York & London.

Kligman, A. M. (1978). Solar damage to the skin. *Drug and Cosmetic Industry*, **123**, 33–113.

Kohn, R. R. & Rollerson, E. (1960). Ageing of human collagen in relation to susceptibility to the action of collagenase. *Journal of Gerontology*, **15**, 10–14.

Korlof, B., Simoni, E., Baryd, I., Lamke, L. O. & Erikson, G. (1972). Radiation-sterilized split skin: a new type of biological wound dressing. *Scandinavian Journal of Plastic and Reconstructive Surgery*, **6**, 126–31.

Kornberg, J., Burns, N. E., Kafesjian, R. & Bartlett, R. M. (1972). Ultrathin silicone polymer membrane: a new synthetic skin substitute. A preliminary study. *Transactions of the American Society for Artificial Internal Organs*, **18**, 39–44.

Kubanyi, A. (1945). Trapianto d'amnion sterile offenuto dal teglio cesareo. *Annali italiani di chirurgia*, **25**, 10–14.

Kumakiri, M., Hashimoto, K. & Willis, I. (1977). Biologic changes due to long-wave ultraviolet radiation on human skin: ultrastructural study. *Journal of Investigative Dermatology*, **69**, 392–400.

Kuno, Y. (1934). *The Physiology of Human Perspiration*. Churchill, London.

Kuokkanen, K. (1972). Replica reflection of normal skin and of skin with disturbed keratinisation. *Acta Dermatovener (Stockholm)*, **52**, 205–10.

Lamke, L. O. (1970). An instrument for estimating evaporation from small skin surfaces. *Scandinavian Journal of Plastic amd Reconstructive Surgery*, **4**, 1–6.

Lamke, L. O. (1971). The influence of different skin grafts on the evaporative water loss from burns. *Scandinavian Journal of Plastic and Reconstructive Surgery*, **5**, 82–6.

Lamke, L. O. & Liljedahl, S. O. (1971). Evaporative water loss from burns, grafts and donor sites. *Scandinavian Journal of Plastic and Reconstructive Surgery*, **5**, 17–22.

Lamke, L. O., Nilsson, G. E. & Reithner, H. L. (1977a). Insensible perspiration from the skin under standardised environmental conditions. *Scandinavian Journal of Clinical Laboratory Investigation*, **37**, 325–31.

Lamke, L. O., Nilsson, G. E. & Reithner, H. L. (1977b). The evaporative water loss from burns and the water-vapour permeability of grafts and artificial membranes used in the treatment of burns. *Burns*, **3**, 159–65.

Langer, K. (1861), translated by Gibson, T. (1978). On the anatomy and physiology of the skin. I. The cleavability of the cutis. *British Journal of Plastic Surgery*, **31**, 3–8. (*from Sitzungsberichte der mathematisch-naturwissenschaftlichen Klasse der Kaiserlichen Akademie der Wissenschaften*, **44**, 19.)

Langer, K. (1862a), translated by Gibson, T. (1978). On the anatomy and physiology of the skin. II. Skin tension. *British Journal of Plastic Surgery*, **31**, 93–106. (From *Sitzungsberichte der mathematisch-naturwissenschaftlichen Klasse der Kaiserlichen Akademie der Wissenschaften*, **45**, 133.)

Langer, K. (1862b), translated by Gibson, T. (1978). On the anatomy and physiology of the skin. III. The elasticity of the cutis. *British Journal of Plastic Surgery*, **31**, 185–99. (From *Sitzungsberichte der mathematisch-naturwissenschaftlichen Klasse der Kaiserlichen Akademie der Wissenschaften*, **45**, 156.)

Langerhans, P. (1868). Über die Nerven der menschlichen Haut. *Virchows Archiv für pathologische Anatomie und Physiologie und für klinische Medizin*, **44**, 325–37.

Lanir, Y. (1976). Biaxial stress relaxation in skin. *Annals of Biomedical Engineering*, **4**, 250–70.

Lanir, Y. (1979). A structural theory for the homogeneous biaxial stress–strain relationships in flat collagenous tissue. *Journal of Biomechanics*, **12**, 423–36.

Lanir, Y. & Fung, Y. C. (1974). Two dimensional mechanical properties of rabbit-skin. II. Experimental results. *Journal of Biomechanics*, **7**, 171–82.

Laurence, E. B., Spargo, D. J. & Thornley, A. L. (1979). Cell proliferation kinetics of epidermis and sebaceous glands in relation to chalone action. *Cell and Tissue Kinetics*, **12**, 615–33.

Lavker, R. M. (1979). Structural alterations in exposed and unexposed aged skin. *Journal of Investigative Dermatology*, **73**, 59–66.

Lefevre, J. (1901). Studies on the thermal conductivity of skin *in vivo* and the variations induced by changes in the surrounding temperature. *Journale de physique* (June).

Le Gros Clark, W. E. & Buxton, L. H. (1938). Studies in nail growth. *British Journal of Dermatology and Syphilis*, **50**, 221–7.

Lendrum, J. & Bowen-Jones, E. (1975). A new dressing for burns. Enclosure in a plasticised polyvinyl chloride sheet. *Burns*, **2**, 86–9.

Levan, N. E., Hyman, C., Freedman, R. & Rohter, F. (1964). Biphasic changes in cutaneous effective blood flow after irradiation. *Journal of Investigative Dermatology*, **43**, 451–2.

Lewis, B. L. (1954). Microscopic studies of foetal and mature nail and surrounding soft tissue. *Archives of Dermatology and Syphilology*, **70**, 732–47.

Lianis, G. (1963). *Constitutive Equations of Viscoelastic Solids under Large Deformation*. Purdue University School of Aeronautical and Engineering Sciences Report No. A. & E.S., 63–65. Purdue University, Lafayette, Indiana.

Lipkin, M. & Hardy, J. D. (1954). Measurement of some thermal properties of human tissues. *Journal of Applied Physiology*, **7**, 212–17.

Loesch, D. (1974). Genetical studies of sole and palmar dermatoglyphics. *Annals of Human Genetics*, **37**, 405–20.

Lorincz, A. L. (1954). Pigmentation. In *Physiology and Biochemistry of the Skin*, ed. S. Rothman, pp. 515–63. University of Chicago Press.

Lund, H. Z. & Sommerville, R. L. (1957). Basophilic degeneration of the cutis. Data substantiating its relationship to prolonged solar exposure. *American Journal of Clinical Pathology*, **27**, 183–90.

Lynfield, Y. L. (1960). Effect of pregnancy on the human hair cycle. *Journal of Investigative Dermatology*, **35**, 323–7.

MacDonald, D. M. & Schmitt, D. (1979). Ultrastructure of the human mucocutaneous end organ. *Journal of Investigative Dermatology*, **72**, 181–6.

McDowall, R. A. W. & Hackett, M. E. J. (1976). The use of lyophilised skin as an initial dressing in partial thickness burns. *Chirurgia Plastica (Berlin)*, **3**, 159–64.

McGregor, I. A. (1965). *Fundamental Techniques of Plastic Surgery*. Williams & Wilkins Co., Baltimore.

Mackie, B. S. & McGovern, V. J. (1958). The mechanism of solar carcinogenesis. *Archives of Dermatology*, **78**, 218–44.

McKnight, J. J. & Guldalian, J. (1974). Laminated collagen film dressing. US Patent Application no. 244 439 (2 April).

MacLachlan, T. B. (1965). A method for the investigation of the strength of the fetal membranes. *American Journal of Obstetrics and Gynecology*, **91**, 309–13.

McLean, A. L. (1919). Bacteriological and other researches. In *Australian Antarctic Expedition 1911–1914, Scientific Report*, ser. C, vol. VII, part 4. Australian Government Publication, Sydney.

MacLeod, T. M. & Frain-Bell, W. (1975). A study of physical light screening agents. *British Journal of Dermatology*, **92**, 149–56.

Maher, V. M. & McCormick, J. J. (1976). Effect of DNA repair on the cytotoxicity and mutagenesis of UV irradiation and of chemical carcinogens in normal and xeroderma pigmentosum cells. In *Biology of Radiation Carcinogenesis*, ed. J. M. Yuhas, R. W. Tennant & J. D. Regan, pp. 129–45.

Maher, V. M., Ouelette, L. M., Curran, R. D. & McCormick, J. J. (1976). Frequency of ultraviolet light induced mutations is higher in xeroderma pigmentosum variant cells than in normal human cells. *Nature (London)*, **261**, 593–5.

Makki, S. (1980). A quantitative method for evaluating the microtopography of the surface of human skin. PhD thesis, University of Strathclyde, Glasgow.

Makki, S., Barbenel, J. C. & Agache, P. (1979). A quantitative method for the assessment of the microtopography of human skin. *Acta Dermatovener (Stockholm)*, **59**, 285–91.

Mali, J. W. H. (1956). The transport of water through the human epidermis. *Journal of Investigative Dermatology*, **27**, 451–69.

Maloney, M. J., Paquette, E. G. & Shansky, A. (1977). The physical properties of fingernails. I. Apparatus for physical measurements. *Journal of the Society of Cosmetic Chemists*, **28**, 415–25.

Malten, K. E. & Thiele, F. A. J. (1973). Evaluation of skin damage. *British Journal of Dermatology*, **89**, 565–9.

Markenscoff, X. & Yannas, I. V. (1979). On the stress–strain relation for skin. *Journal of Biomechanics*, **12**, 127–9.

Marks, F. (1975). Isolation of an endogenous inhibitor of epidermal DNA synthesis (G_1 chalone) from pig skin. *Hoppe-Seyler's Zeitschrift für physiologische Chemie (Berlin)*, **356**, 1989–92.

Marks, R. & Dawber, R. P. R. (1971). Skin surface biopsy: an improved technique for the examination of the horny layer. *British Journal of Dermatology*, **84**, 117–23.

Marks, R. & Pearse, A. D. (1975). Surfometry. *British Journal of Dermatology*, **92**, 651–7.

Marshall, D. R. (1965). The clinical and pathological effects of prolonged solar exposure. I. The association with ageing of the skin. *Australian and New Zealand Journal of Surgery*, **4**, 161–72.

Masson, P. (1926). Les naevi pigmentaires, tumeurs nerveuses: (I and II). *Annales d'anatomie pathologiques et d'anatomie normale médico-chirurgicale*, **3**, 417–53, 657–96.

Matoltsy, A. G. (1967). The envelope of epidermal horny cells. *Congress on Investigative Dermatology*, **13**, 1014.

Matoltsy, A. G., Downes, A. M. & Sweeney, T. M. (1968). Studies of the epidermal water barrier. II. Investigation of the chemical nature of the water barrier. *Journal of Investigative Dermatology*, **50**, 19–26.

Matoltsy, A. G., Schragger, A. & Matoltsy, M. N. (1962). Observations on regeneration of the skin barrier. *Journal of Investigative Dermatology*, **38**, 251–3.

Matsumoto, S. (1913). Über eine eigentümliche Pigment Verteilung an den voigtschen Linien. *Archiv für Dermatologie und Syphilis*, **118**, 157–61.

Meara, R. H. (1964). Atrophic striae following topical fluocinolone therapy. *British Journal of Dermatology*, **76**, 481–2.

Meigel, W. N. & Weber, L. (1976). Hautalterung: ein Problem der Kollagenpolymorphie. *Zeitschrift für Gerontologie*, **2**, 377–86.

Melski, J. W., Tanenbaum, L., Parrish, J. A., Fitzpatrick, T. B. & Bleich, H. L. (1977). Oral methoxsalen photochemotherapy for the treatment of psoriasis: a co-operative clinical trial. *Journal of Investigative Dermatology*, **68**, 328–35.

Middleton, J. D. (1968). The mechanism of water binding in stratum corneum. *British Journal of Dermatology*, **80**, 437–50.

Middleton, J. D. (1969). The mechanism of action of surfactants on the water binding properties of isolated stratum corneum. *Journal of the Society of Cosmetic Chemists*, **20**, 399–403.

Middleton, J. D. & Allen, B. M. (1973). The influence of temperature and humidity on stratum corneum and its relation to skin chapping. *Journal of the Society of Cosmetic Chemists*, **24**, 239–43.

Miller, L. K. & Irving, L. (1962). Local reactions to air cooling in an Eskimo population. *Journal of Applied Physiology*, **17**, 449–55.

Miller, R. M. & Coger, R. W. (1979). Skin conductance conditioning with dishydrotic eczema patients. *British Journal of Dermatology*, **101**, 435–40.

Millington, P. F., Gibson, T., Evans, J. H. & Barbenel, J. C. (1971). Structural and mechanical properties of connective tissues. In *Advances in Biomechanical Engineering*, vol. 1, ed. R. M. Kenedi, pp. 189–248. Academic Press, New York & London.

Minns, R. J., Soden, P. D. & Jackson, D. S. (1973). The role of the fibrous components and ground substance in the mechanical properties of biological tissues: a preliminary investigation. *Journal of Biomechanics*, **6**, 153–65.

Mishima, Y. (1971). Introduction to diagnostic histopathology. *Acta Dermatovener, Stockholm*, **51**, 16–20.

Molokhia, M. M. & Portnoy, B. (1973). Trace elements and skin pigmentation. (Comment.) *British Journal of Dermatology*, **89**, 207–9.

Monash, S. & Blank, H. (1958). Location and re-formation of the epithelial barrier to water vapour. *Archives of Dermatology and Syphilogy*, **78**, 710–4.

Moncrief, J. A. & Mason, A. D. (1962). Water vapour loss in the burned patient. *Surgical Forum*, **13**, 38–41.

Montagna, W., Ellis, R. A. & Silver, A. F. (eds.) (1963). *Advances in the Biology of Skin*, vol. 4. Pergamon Press, New York.

Montagna, W. & Parakkal, P. E. (1974). *The Structure and Function of Skin*, 3rd edn. Academic Press, New York & London.

Moritz, A. R. & Henriques, F. C. (1947). Studies of thermal injury. II. The relative importance of time and surface temperatures in the causation of cutaneous burns. *American Journal of Pathology*, **23**, 695–720.

Morris, D. M., Hall, G. M. & Elias, E. G. (1979). Porcine heterograft dressings for split thickness graft donor sites. *Surgery, Gynecology and Obstetrics*, **149**, 893–4.

Moynahan, E. J. & Engel, C. E. (1962). Photomacrography of the normal skin. *Medical and Biological Illustration*, **12**, 72–82.

Muehrcke, R. C. (1956). The finger nails in chronic hypoalbumaemia. *British Medical Journal*, **i**, 1327–8.

Mukherjee, D. P. (1969). Viscoelastic properties of elastin. PhD thesis, Massachusetts Institute of Technology, Cambridge, USA.

Nathan, P., Law, E. J., MacMillan, B. G., Murphy, D. F., Rouel, S. H., D'Andrea, M. J. & Abrahams, R. A. (1976). A new biomaterial for the control of infection in the burn wound. *Transactions of the American Society for Artificial Internal Organs*, **22**, 30–41.

Nemetschek, T., Jonak, R., Nemetschek-Gansler, H. & Reidl, H. (1978). Determination of changes in the large periodic structure of collagen. *Zeitschrift für Naturforschung*, **33c**, 928–36.

Newburgh, L. M. (ed.) (1949). *The Physiology of Heat Regulation*. Saunders, Philadelphia.

Nicholls, S., King, C. S. & Marks, R. (1978). Short term effects of emollients and a bath oil on the stratum corneum. *Journal of the Society of Cosmetic Chemists*, **29**, 617–24.

Nicholls, S. & Marks, R. (1977). Novel techniques for the estimation of intracorneal cohesion *in vivo*. *British Journal of Dermatology*, **96**, 595–602.

Nilsson, G. E. & Oberg, P. A. (1979). Measurements of evaporative water loss: methods and clinical application. In *Non-invasive Physiological Measurements*, vol. 1, ed. P. Rolfe, chapt. 12. Academic Press, New York & London.

North, J. F. & Gibson, F. (1978). Volume compressibility of human abdominal skin. *Journal of Biomechanics*, **11**, 203–7.

Oberste-Lehn, H. (1962). Dermo-epidermal interface. *Archives of Dermatology*, **86**, 770–8.

Okubo, T. & Sano, S. (1973). Functional aspects of dermo-epidermal junction. *Acta Dermatovener (Stockholm), Supplement*, **73**, 121–8.

Orentriech, N., Markofsky, J. & Vogelman, J. H. (1979). The effect of ageing on the rate of linear nail growth. *Journal of Investigative Dermatology*, **3**, 126–30.

Orfanos, C., Christenhusz, R. & Mahrle, G. (1969). Die normale und psoriatische Hautoberflache vergleichende Beobachtungen mit dem Raster-Elektronenmikroskop. *Archiv für klinische und experimentelle Dermatologie*, **235**, 284–94.

Orth, J. (1887). *Lehrbuch der spezialen pathologische Anatomie*. Hirschwald, Berlin.

Packman, E. W. & Gans, E. H. (1978). The panel study as a scientifically controlled investigation: moisturisers and superficial facial lines. *Journal of the Society of Cosmetic Chemists*, **29**, 91–8.

Palmes, E. D. (1948). An apparatus and method for the continuous measurement of evaporative water loss from human subjects. *Review of Scientific Instruments*, **9**, 711–17.

Parakkal, P. F. & Matoltsy, A. G. (1964). A study of the differentiation products of the hair follicle cells with electron microscopy. *Journal of Investigative Dermatology*, **42**, 23–34.

Park, A. C. & Baddiel, C. B. (1972a). Rheology of stratum corneum. I. A molecular interpretation of the stress–strain curve. *Journal of the Society of Cosmetic Chemists*, **23**, 3–12.

Park, A. C. & Baddiel, C. B. (1972b). Rheology of stratum corneum. II. A physicochemical investigation of factors influencing the water content of the corneum. *Journal of the Society of Cosmetic Chemists*, **23**, 13–21.

Park, A. C. & Baddiel, C. B. (1972c). The effect of saturated salt solutions on the elastic properties of stratum corneum. *Journal of the Society of Cosmetic Chemists*, **23**, 471–9.

Park, G. B., Courtney, J. M., McNair, A. & Gaylor, J. D. S. (1978). The design and evaluation of a burn wound covering. *Engineering in Medicine*, **7**, 11–15.

Park, G. B. & Mollison, D. S. (1978). *In vitro* bacteriological study on film burn wound coverings. *Burns*, **6**, 96–104.

Park, G. B., Mollison, D. S., Courtney, J. M., Gaylor, J. D. S. & McNair, A. (1977). Polymeric films as burn wound coverings. *Proceedings of the European Society for Artificial Internal Organs*, **3**, 154–8.

Parmley, T. H. & Seeds, A. E. (1970). Fetal skin permeability to isotopic water (THO) in early pregnancy. *American Journal of Obstetrics and Gynecology*, **108**, 128–31.

Parrish, J. A., Anderson, R. R., Urbach, F. & Pitts, D. (1978). *UV-A*. Plenum Press, New York.

Parrish, J. A., Ying, C. Y., Pathak, M. A. & Fitzpatrick, T. B. (1974). Erythemogenic properties of long-wave ultraviolet light. In *Sunlight and Man: Normal and Abnormal Photobiologic Responses*, ed. M. A. Pathak, M. Harber, M. Seiji & A. Kukita, pp. 131–41. University of Tokyo Press.

Partridge, S. M. (1962). Elastin. *Advances in Protein Chemistry*, **17**, 227–302.

Pathak, M. A. & Kramer, D. M. (1969). Photosensitization of skin, *in vivo*, by Furocoumarins (Psoralens). *Biochimica et Biophysica Acta*, **195**, 197–206.

Pathak, M. A., Riley, F. L. & Fitzpatrick, T. B. (1962). Melanogenesis in human skin following exposure to ultraviolet and visible light. *Journal of Investigative Dermatology*, **39**, 435–46.

Pecoraro, V., Astore, I. & Barman, J. M. (1971). Growth rate and hair density of the human axilla. *Journal of Investigative Dermatology*, **56**, 362–5.

Penrose, L. S. & Loesch, D. (1969). Dermatoglyphic sole patterns: a new attempt at classification. *Human Biology*, **41**, 427–8.

Peter, J. & Wyndham, C. H. (1966). Activity of the human eccrine sweat gland during exercise in a hot humid environment before and after acclimatization. *Journal of Physiology*, **187**, 583–94.

Pethig, R. (1979). *Dielectric and Electronic Properties of Biological Materials*. Wiley, New York.

Pierard, G. E. & Lapiere, C. H. M. (1976). Skin in dermatosparaxis: dermal micro-architecture and biomechanical properties. *Journal of Investigative Dermatology*, **66**, 2–7.

Pigeon, J. (1960). Treatment of second-degree burns with amniotic membranes. *Canadian Medical Association Journal*, **83**, 844–5.

Pinkus, F. (1910). Development of the integument. In *Manual of Embryology*, ed. F. Keibel & F. P. Mall, pp. 243–91. Lippincott, Philadelphia.

Pinkus, F. (1927). Normale Anatomie der Haut. In *Handbuch der Haut und Geschlechtskrankheiten*, ed. J. Jadassohn, vol. 1, pp. 1–110. Springer, Berlin.

Pinnell, S. R., Eisen, A. Z., Graham Smith, J., Bauer, E. A., Fleischmajer, R., Silbert, J. & Uitto, J. J. (1979). Connective tissue diseases and aging. *Journal of Investigative Dermatology*, **73**, 485–90.

Plewig, G. & Marples, R. (1969). Regional differences of cell sizes in the human stratum corneum: I. *Journal of Investigative Dermatology*, **54**, 13–18.

Plewig, G. & Marples, R. (1970). Regional differences of cell sizes in the human stratum corneum. II. *Journal of Investigative Dermatology*, **54**, 19–23.

Pochi, P. E., Strauss, J. S. & Downing, D. T. (1979). Age related changes in sebaceous gland activity. *Journal of Investigative Dermatology*, **73**, 108–11.

Poidevin, L. O. S. (1959). Striae gravidarum: their relation to adrenal cortical hyperfunction. *Lancet*, **ii**, 436–9.

Polishuk, W. Z., Kohane, S. & Hadar, A. (1964). Fetal weight and membrane tensile strength. *American Journal of Obstetrics and Gynecology*, **88**, 247–50.

Porter, J. & Fouweather, C. (1975). An appraisal of human head hair in forensic evidence. *Journal of the Society of Cosmetic Chemists*, **26**, 299–313.

Potten, C. S. (1975). Epidermal cell production rates. *Journal of Investigative Dermatology*, **65**, 488–500.

Potten, C. S. & Allen, T. D. (1976). A model implicating the Langerhans cell in keratinocyte proliferation control. *Differentiation*, **5**, 443–7.

Prall, J. K. (1973). Instrumental evaluation of the effects of cosmetic products on skin surfaces with particular reference to smoothness. *Journal of the Society of Cosmetic Chemists*, **24**, 693–707.

Price, V. H. (1975). Testosterone metabolism in the skin. *Archives of Dermatology*, **111**, 1496–502.

Pringle, R. (1963). Amnion implantation in peripheral vascular disease. *Lancet*, **i**, 77–8.

Pruitt, B. A. & Silverstein, P. (1971). Methods of resurfacing denuded skin areas. *Transplantation Proceedings*, **3**, 1537–45.

Quevedo, W. C. & Fleischmann, R. D. (1980). Developmental biology of mammalian melanocytes. *Journal of Investigative Dermatology*, **75**, 116–20.

Raekallio, J. & Seligman, A. M. (1964). Acute reaction to arterial adhesive in healing skin wounds. *Journal of Surgical Research*, **4**, 124–7.

Rajka, G. & Thune, P. (1976). The relationship between the course of psoriasis and transepidermal water loss, photoelectric plethysmography and reflex photometry. *British Journal of Dermatology*, **94**, 253–61.

Ramsay, C. A. (1979). Skin responses to ultraviolet radiation in contact photodermatitis due to Fentichlor. *Journal of Investigative Dermatology*, **72**, 99–102.

Ramsay, C. A. & Challoner, A. V. J. (1976). Vascular changes in human skin after ultraviolet irradiation. *British Journal of Dermatology*, **94**, 487–93.

Ranu, H. S. (1981). Effects of fractionated doses of X-irradiation on the mechanical properties of skin: a long term study: In *Bioengineering and the Skin*, ed. R. Marks & P. A. Payne, pp. 15–21. MTP Press, Lancaster.

Ranu, H. S., Burlin, T. E. & Hutton, W. C. (1975). The effects of X-radiation on the mechanical properties of skin. *Physics in Medicine and Biology*, **20**, 96–105.

Rao, R. B. (1972). PhD thesis, Madras, India. (Quoted by Minns *et al.*, 1973.)

Rebell, G. & Kirk, D. (1962). Patterns of eccrine sweating in the human axilla. In *Advances in the Biology of Skin*, vol. 3, ed. W. Montagna, R. A. Ellis & A. F. Silver, pp. 108–26. Pergamon Press, New York.

Reid, W. H. (1974). Care of the burned hand. *Hand*, **6**, 163–5.

Reller, H. H. (1964). Factors affecting axillary sweating. *Journal of the Society of Cosmetic Chemists*, **15**, 99–110.

Rhodin, J. A. G. (1973). Structure, pattern and shape of the blood vessels of the skin. In *The Physiology and Pathology of the Skin*, vol. 2, ed. A. Jarrett, pp. 595–632. Academic Press, New York & London.

Ridge, M. D. & Wright, V. (1964). The description of skin stiffness. *Biorheology*, **2**, 67–74.

Rigby, B. J., Hirai, R., Spikes, J. D. & Eyring, H. (1959). The mechanical properties of rat tail tendon. *Journal of General Physiology*, **43**, 265–83.

Robbins, J. H. & Moshell, A. N. (1979). DNA repair processes protect human beings from premature solar skin damage: evidence from studies on xeroderma pigmentosum. *Journal of Investigative Dermatology*, **73**, 102–7.

Roberts, D. F. (1977). Human pigmentation: its geographical and facial distribution and biological significance. *Journal of the Society of Cosmetic Chemists*, **28**, 329–42.

Roberts, M. (1976). The role of the skin bank. *Annals of the Royal College of Surgeons of England*, **58**, 70–4.

Robertson, W. van B. (1964). Metabolism of collagen in mammalian tissue. *Biophysical Journal (New York)*, **4** (*Supplement*), 93–114.

Robson, M. C. & Krizek, T. J. (1974). Clinical experiences with amniotic membranes as a temporary biologic dressing. *Connecticut Medicine*, **38**, 449–53.

Rockwell, R. J. (1971). The current status of laser applications in medicine and biology. *CRC Critical Reviews in Bioengineering*, **1**, 49–67.

Rodrigo, F. & Cotta-Pereira, G. (1979). Connective fibres in the dermo-epidermal anchorage. *Dermatologica*, **158**, 13–23.

Roeder, F. (1934). Die Miessung der Warmeleitzahl der menschlichen Haut und ihre Veranderlichkeit. *Zeitschrift für Biologie*, **95**, 164–8.

Rogers, B. O. (1951). Guide and bibliography for research into skin homograft problem. *Plastic and Reconstructive Surgery*, **7**, 169–207.

Rolfe, P. (ed.) (1979). *Non-invasive Physiological Measurements*. Academic Press, New York & London.

Rook, A. (1965). Some chemical influences on hair growth and pigmentation. *British Journal of Dermatology*, **77**, 115–29.

Rosario, R., Mark, G. J., Parrish, J. A. & Mihm, M. (1979). Histological changes produced in skin by equally erythemogenic doses of UV-A, UV-B, UV-C and UV-A with psoralens. *British Journal of Dermatology*, **101**, 299–308.

Rosencrants, C. Z. (1930). A high-sensitivity absolute-humidity recorder. *Industrial and Engineering Chemistry (Analytical Edition)*, **2**, 129–34.

Rosenthal, S. R. (1977). Histamine as the chemical mediator for cutaneous pain. *Journal of Investigative Dermatology*, **69**, 98–105.

Ross, R. (1973). The elastic fiber. *Journal of Histochemistry and Cytochemistry*, **21**, 199–208.

Ross, R. & Bornstein, P. (1969). Elastic fibre. I. Separation and partial characterisation of its macro-molecular components. *Journal of Cell Biology*, **40**, 366–81.

Rothman, S. (1954). *Physiology and Biochemistry of Skin*. University of Chicago Press.

Rowell, L. B. (1977). Reflex control of the cutaneous vasculature. *Journal of Investigative Dermatology*, **69**, 154–66.

Ryan, J. J. & Kurban, A. K. (1970). New vessel growth in the adult skin. *British Journal of Dermatology*, **82**, Supplement 5, 92–8.

Saitoh, M., Uzuka, M. & Sakamoto, M. (1970). The human hair cycle. *Journal of Investigative Dermatology*, **54**, 65–81.

Salter, D. C. (1981a). Alternating current electrical properties of human skin measured in vivo. In *Bioengineering and the Skin*, ed. R. Marks & P. A. Payne, pp. 267–74. MTP Press, Lancaster.

Salter, D. C. (1981b). Studies in the measurement, form and interpretation of some electrical properties of normal and pathological skin *in vivo*. DPhil thesis, University of Oxford, England.

Sammon, P. D. (1972). Nail formation and some nail disorders. *Journal of the Society of Cosmetic Chemists*, **23**, 405–13.

Sanders, R. (1973). Torsional elasticity of human skin in vivo. *Pflügers Archiv*, **342**, 255–60.

Sarkany, I. (1962). A method for studying the microtopography of the skin. *British Journal of Dermatology*, **74**, 254–9.

Sarkany, I. & Gaylarde, P. (1968). A method for demonstration of sweat gland activity. *British Journal of Dermatology*, **80**, 601–5.

Sato, L. & Dobson, R. L. (1970). Regional and individual variations in the function of the human sweat gland. *Journal of Investigative Dermatology*, **54**, 443–9.

Schade, H. (1920). *Die physikalische Chemie in der inneren Medizin*. T. Steinkopff, Dresden & Leipzig.

Schaumann, B. & Alter, M. A. (1976). *Dermatoglyphics in Medical Disorders*. Springer, New York & Berlin.

Schellander, F. A. & Headington, J. T. (1974). The stratum corneum: some structural and functional correlates. *British Journal of Dermatology*, **91**, 507–15.

Scheuplein, R. J. (1965). Mechanism of percutaneous absorption. *Journal of Investigative Dermatology*, **45**, 334–46.

Scheuplein, R. J. (1966). Analysis of permeability data for the case of parallel diffusion pathways. *Biophysical Journal*, **6**, 1–17.

Scheuplein, R. J. & Blank, I. H. (1971). Permeability of the skin. *Physiological Reviews*, **51**, 702–47.

Scheuplein, R. J. & Morgan, L. J. (1967). Bound-water in keratin membranes, measured by a microbalance technique. *Nature (London)*, **214**, 456–8.

Scheuplein, R. & Ross, L. (1970). Effects of surfactants and solvents on the permeability of the epidermis. *Journal of the Society of Cosmetic Chemists*, **21**, 853–73.

Schweizer, J. & Marks, F. (1977). A developmental study of the distribution and frequency of Langerhans cells in relation to formation of patterning in mouse tail epidermis. *Journal of Investigative Dermatology*, **69**, 198–204.

Shahidullah, M., Raffee, E. J. & Frain-Bell, W. (1968). Insensible water loss in dermatitis. *Journal of Dermatology*, **78**, 589–94.

Shaw, L. A., Messer, A. C. & Weiss, S. (1929). Cutaneous respiration in man. I. Factors affecting the rate of carbon dioxide elimination and oxygen absorption. *American Journal of Physiology*, **90**, 107–18.

Sheppard, R. H. & Meema, H. E. (1967). Skin thickness in endocrine disease. A roentgenographic study. *Annals of Internal Medicine*, **66**, 531–9.

Shuck, J. M., Bedeau, G. W. & Thomas, P. R. (1972). Homograft skin for the early management of difficult wounds. *Journal of Trauma*, **12**, 215–22.

Shuster, S. (1963). Graded sweat-duct occlusion: a technique for studying sweat-gland function. *Clinical Science*, **20**, 89–95.

Shuster, S. & Bottoms, E. (1963). Senile degeneration of skin collagen. *Clinical Science*, **25**, 487–91.

Silver, A., Montagna, W. & Karacan, I. (1964). Age and sex differences in spontaneous adrenergic and cholinergic sweating. *Journal of Investigative Dermatology*, **43**, 255–65.

Silvers, W. K. (1957). A histological and experimental approach to determine the relationship between gold impregnated dendritic cells and melanocytes. *American Journal of Anatomy*, **100**, 225–40.

Silvers, W. K. (1979). *The Coat Colors of Mice: A Model for Mammalian Gene Action and Interaction*. Springer, New York & Berlin.

Silvetti, A. N., Cotten, D., Byrne, R. J., Berrian, J. H. & Menendez, A. F. (1957). Preliminary studies of bovine embryo skin grafts. *Transplanation Bulletin*, **4**, 25–9.

Silvetti, A. N., Teresi, M., Luangkesorn, P., Khan, T., Locke, R. & Mass, A. (1979). Bovine embryonic skin allografts in treatment of skin ulcers and burns in man. *Transplantation Proceedings*, **11**, 1512–13.

Simpson, N. B., Bowden, F. E., Forster, R. A. & Cunliffe, W. J. (1979). The effect of topically applied progesterone on sebum excretion rate. *British Journal of Dermatology*, **100**, 687–92.

Sinclair, D. (1967). *Cutaneous Sensations*. Oxford University Press.

Singer, E. J. & Vinson, L. J. (1966). The water-binding properties of skin. *Proceedings of the Scientific Section of the Toilet Goods Association*, **46**, 29–31.

Skog, E. & Wahlberg, J. E. (1964). A comparative investigation of the percutaneous absorption of metal compounds in the guinea pig by means of the radioactive isotopes: ^{51}Cl, ^{58}Co, ^{65}Zn, ^{110m}Ag, ^{115m}Cd, ^{203}Hg. *Journal of Investigative Dermatology*, **43**, 187–92.

Sliney, D. & Wolbarsht, M. (1981). *Safety with Lasers and Other Optical Sources*, 2nd ed. Plenum Press, New York.

Smeenk, G. (1968). The influence of detergents on the skin. *Archiv für klinische und experimentelle Dermatologie*, **235**, 180–91.

Smith, K. (1977). The Haarscheibe. *Journal of Investigative Dermatology*, **69**, 68–74.

Sneddon, I. B. (1976). Atrophy of the skin: the clinical problems. *British Journal of Dermatology*, **94**, *Supplement 12*, 121–3.

Snyder, R. W. (1972). Large deformations of isotropic biological tissue. *Journal of Biomechanics*, **5**, 601–6.

Soong, T. T. & Huang, W. N. (1973). A stochastic model for biological tissue elasticity in simple elongation. *Journal of Biomechanics*, **6**, 451–8.

Spruit, D. (1967). Measurement of the water vapour loss from human skin by a thermal conductivity cell. *Journal of Applied Physiology*, **23**, 994–7.

Spruit, D. & Malten, K. E. (1966). The regeneration rate of the water vapour loss of heavily damaged skin. *Dermatologica*, **132**, 115–23.

Spruit, D. & Reynen, A. Th. A. (1972). Pattern of sweat gland activity on the forearm, after pharmacologic stimulation. *Acta Dermatovener (Stockholm)*, **52**, 129–35.

Squier, C. A. (1980). The stretching of mouse skin *in vivo*: effect on epidermal proliferation and thickness. *Journal of Investigative Dermatology*, **74**, 68–71.

Staats, W. F., Foskett, L. W. & Jensen, H. P. (1965). Infra-red absorption hygrometer. In *Humidity and Moisture*, ed. A. Wexter & R. E. Ruskin, vol. 1. Reinhold, New York.

Stankler, L. (1977). Histological findings in clinically normal palmar skin of patients with psoriasis. *British Journal of Dermatology*, **97**, 131–4.

Stark, H. L. (1977). Directional variation in the extensibility of human skin. *British Journal of Plastic Surgery*, **30**, 105–14.

Stefanovic, D. C. (1976). Corticosteroid-induced atrophy of the epidermis. In *Mechanisms of Topical Corticosteroid Activity*, ed. L. C. Wilson & R. Marks, pp. 97–114. Churchill Livingstone, Edinburgh.

Steinmetz, M. A. & Adams, T. (1981). Epidermal water and electrolyte content and the thermal, electrical and mechanical properties of skin. In *Bioengineering and the Skin*, ed. R. Marks & P. A. Payne, pp. 197–213. MTP Press, Lancaster.

Stenstrom, S., Bergman, F. & Bergman, S. (1972). Wound healing with ordinary adhesive tape. A clinical and experimental study. *Scandinavian Journal of Plastic and Reconstructive Surgery*, **6**, 40–6.

Sterling, J. A. (1956). Use of amniotic membranes to cover surface defects due to flame burns. *American Journal of Surgery*, **91**, 940–2.

Stern, R. S., Zeirler, S. & Parrish, J. A. (1980). Skin carcinoma in patients with psoriasis treated with topical tar and artificial ultraviolet radiation. *Lancet*, **ii**, 732–5.

Sternberg, T. H., Levan, P. & Wright, E. T. (1961). The hydrating effect of pregnenalone acetate on the human skin. *Current Therapeutic Research*, **3**, 469–71.

Stevens, J. C. & Jones, N. B. (1977). A review of apparatus for investigating the mechanical properties of soft animal tissue. *Engineering in Medicine*, **6**, 112–19.

Stevenson, K. L. (1977). Suturing. *Surgical Clinics of North America*, **57**, 863–73.

Stingl, G., Katz, S. I., Green, I. & Shevach, E. M. (1980). The functional role of Langerhans cells. *Journal of Investigative Dermatology*, **74**, 315–18.

Stoll, A. M. (1977). Thermal properties of human skin related to non-destructive measurement of epidermal thickness. *Journal of Investigative Dermatology*, **69**, 328–32.

Stoll, A. M. & Chianta, M. A. (1982). *Interface Material Skin Temperatures Related to Epidermal Thickness Measurement*. Report No. NADC-82020-60. Department of the Navy, Washington, DC.

Stoll, A. M. Chianta, M. A. & Piergallini, J. R. (1979). Thermal conductance effects in human skin. *Aviation, Space and Environmental Medicine*, **50**, 778–87.

Stoll, A. M., Chianta, M. A. & Piergallini, J. R. (1982). *Prediction of Threshold Pain*

Skin Temperature from Thermal Properties of Materials in Contact. Report No. NADC-82019-60. Department of the Navy, Washington, DC.

Stoll, A. M. & Greene, L. C. (1959). Relationship between pain and tissue damage due to thermal injury. *Journal of Applied Physiology*, **14**, 373–82.

Stoughton, R. B. & Rothman, S. (1959). *The Human Integument*. Publication No. 54. AAAS, Washington, DC.

Surwillo, W. W. (1969). Statistical distribution of volar skin potential level in attention and the effects of age. *Psychophysiology*, **6**, 13–16.

Szabo, G. (1962). The number of eccrine sweat glands in human skin. In *Advances in the Biology of Skin*, vol. 3, ed. W. Montagna, R. A. Ellis & A. F. Silver, pp. 1–5. Pergamon Press, New York.

Szakall, A. (1951). Forschung und Gesunderhaltung der Haut. *Fette, Seifen, Anstrich-mittel*, **53**, 399–403.

Szigeti, M., Monnier, G., Jacotot, B., Navarro, N. & Robert, L. (1972). Distribution of ingested ^{14}C-cholesterol in the macromolecular fractions of rat connective tissues. *Connective Tissue Research*, **1**, 145–52.

Tagami, H., Masatoshi, O., Iwatsuki, K., Kanamaru, Y., Yamada, M. & Ichijo, B. (1980). Evaluation of the skin surface hydration, *in vivo*, by electrical measurement. *Journal of Investigative Dermatology*, **75**, 500–7.

Takayasu, S., Wakimoto, H., Itami, S. & Sano, S. (1980). Activity of testosterone 5α-reductase in various tissues of human skin. *Journal of Investigative Dermatology*, **74**, 187–91.

Tamaki, K., Stingl, G. & Katz, S. T. (1980). The origin of Langerhans cells. *Journal of Investigative Dermatology*, **74**, 309–10.

Tavis, M. J., Harvey, J. H., Thornton, J. W., Bartlett, R. H. & Woodroof, E. A. (1975). Modified collagen membrane as a skin substitute: preliminary studies. *Journal of Biomedical Materials Research*, **9**, 285–301.

Thiele, F. A. J. & Malten, K. E. (1973). Evaluation of skin damage. I. Skin resistance measurements with alternating current (impedance measurements). *British Journal of Dermatology*, **89**, 373–82.

Thiele, F. A. J. & Senden, K. G. (1966). Relation between skin temperature and insensible perspiration of the human skin. *Journal of Investigative Dermatology*, **47**, 307–12.

Thomas, E. N., Nordquist, R. E., Scott, J. R. & Everett, M. A. (1964). Ultrastructural changes in stratum corneum induced by ultraviolet light. *Journal of Investigative Dermatology*, **43**, 301–17.

Thomas, F. & Baert, H. (1967). The longitudinal striation of the human nails as means of identification. *Journal of Forensic Medicine*, **14**, 113–17.

Thomas, P. E. & Korr, I. M. (1957). Relationship between sweat gland activity and electrical resistance of skin. *Journal of Investigative Dermatology*, **10**, 505–10.

Thompson, M. W. & Bandler, E. (1973). Finger pattern combinations in normal individuals and in Down's Syndrome. *Human Biology*, **45**, 563–70.

Thorbecke, G. J., Silberberg-Sinakin, I. & Flotte, T. J. (1980). Langerhans cells as macrophages in skin and lymphoid organs. *Journal of Investigative Dermatology*, **75**, 32–43.

Thornley, A. L. & Laurence, E. B. (1976). The specificity of epidermal chalone action: the results of *in vivo* experimentation with two purified skin extracts. *Developmental Biology*, **51**, 10–22.

Toews, G. B., Bergstresser, P. R. & Streilen, J. W. (1980). Langerhans cells: sentinels of skin-associated lymphoid tissue. *Journal of Investigative Dermatology*, **75**, 78–82.

Tong, P. & Fung, Y. C. (1976). Stress–strain relationship for skin. *Journal of Biomechanics*, **9**, 649–57.

Tosti, A., Compagno, G., Fazzini, M. L. & Villardita, S. (1977). A ballistometer for the study of the plasto-elastic properties of skin. *Journal of Investigative Dermatology*, **69**, 315–17.

Townsend, P. L. G. (1977). The quest for cheap and painless donor-site dressing. *Burns*, **2**, 82–5.

Tregear, R. T. (1966). *Physical Function of Skin*. Academic Press, New York & London.

Trelford, J. D., Hanson, F. W. & Anderson, D. G. (1975). Wound healing and the amniotic membrane. *Journal of Medicine*, **6**, 383–8.

Trelford, J. D. & Trelford-Sauder, M. (1979). The amnion in surgery, past and present. *American Journal of Obstetrics and Gynecology*, **134**, 833–45.

Tsuchida, Y. (1979). Rate of skin blood flow in various regions of the body. *Plastic and Reconstructive Surgery*, **64**, 505–8.

Uitto, J. (1979). Biochemistry of the elastic fibers in normal connective tissues and its alterations in diseases. *Journal of Investigative Dermatology*, **72**, 1–10.

Uitto, J. & Lichtenstein, J. R. (1976). Defects in the biochemistry of collagen diseases of connective tissue. *Journal of Investigative Dermatology*, **66**, 59–79.

Underwood, E. J. (1971). *Trace Elements in Human and Animal Nutrition*, 3rd edn. Academic Press, New York & London.

Unna, P. G. (1896). The histopathology of the diseases of the skin. (Quoted by Marshall, 1965.)

Uno, H. (1977). Sympathetic innervation of the sweat glands and piloarrector muscles of macaques and human beings. *Journal of Investigative Dermatology*, **69**, 112–20.

Uttley, M. (1972). Measurement and control of perspiration. *Journal of the Society of Cosmetic Chemists*, **23**, 23–43.

Van Duzee, B.F. (1978). The influence of water content, chemical treatment and temperature on the rheological properties of stratum corneum. *Journal of Investigative Dermatology*, **71**, 140–4.

Van Winkle, W., Hastings, J. C., Barker, E., Hines, D. & Nichols, W. (1975). Effect of suture materials on healing skin wounds. *Surgery, Gynecology and Obstetrics*, **140**, 7–12.

Verbov, J. (1970). Clinical significance and genetics of epidermal ridges: a review of dermatoglyphics. *Journal of Investigative Dermatology*, **54**, 261–71.

Vermorken, A. J. M., Goos, C. M. & Roelops, H. M. (1980). A method for the evaluation of the local antiandrogenic action of 5α-reductase inhibitors on human skin. *British Journal of Dermatology*, **102**, 695–701.

Veronda, D. R. & Westman, R. A. (1970). Mechanical characterisation of skin: finite deformations. *Journal of Biomechanics*, **3**, 111–24.

Viidik, A. (1973). Functional properties of collagenous tissues. *International Review of Connective Tissue Research*, **6**, 127–215.

Vinson, L. J., Kochler, W. R., Lehman, M. D., Masurat, I. & Singer, E. J. (1964). *Basic Studies in Percutaneous Absorption*. Semi-annual Report No. 7. Army Chemical Centre, Edgewood, Md. Sec. 3.

Vlasblom, D. C. (1967). Skin elasticity. PhD thesis, University of Utrecht, The Netherlands.

Vogel, H. G. (1972). Influence of age, treatment with corticosteroids and strain rate on mechanical properties of rat skin. *Biochimica et Biophysica Acta*, **286**, 79–83.

Vogel, H. G. (1973). Stress relaxation in rat skin after treatment with hormones. *Journal of Medicine*, **4**, 19–27.

Vogel, H. G. (1974). Correlation between tensile strength and collagen in rat skin. Effect of age and cortisol treatment. *Connective Tissue Research*, **2**, 177–82.

Vogel, H. G. (1975). Collagen and mechanical strength in various organs of rats treated with D-penicillamine or amino-acetylnitrile. *Connective Tissue Research*, **3**, 237–44.

Vogel, H. G. (1976*a*). Tensile strength, relaxation and mechanical recovery in rat skin as influenced by maturation and age. *Journal of Medicine*, **7**, 177–88.

Vogel, H. G. (1976*b*). Age depending changes in mechanical and biochemical parameters of rat skin. *Aktuelle Gerontologie*, **6**, 477–87.

Vogel, H. G. (1977). Strain of rat skin at constant load (creep experiments): influence of age and desmotropic agents. *Gerontology*, **23**, 77–86.

Vogel, H. G. & Hilgner, W. (1979). The 'step' phenomenon as observed in animal skin. *Journal of Biomechanics*, **12**, 75–81.

Vollum, D. I. (1971). Skin markings in negro children from the West Indies. *British Journal of Dermatology*, **85**, 260–3.

von Gierke, H. E. (1962). Biomechanics of impact injury. In *Impact, Acceleration and Stress*, ed. R. Hume. NAS-NRC, Publication No. 977. National Research Council, Washington, DC.

Voorhees, J. J. & Duell, E. A. (1971). Psoriasis as a possible defect of the adenyl-cyclase–cyclic AMP cascade. *Archives of Dermatology*, **104**, 352–8.

Vossoughi, J. & Vaishnav, R. N. (1979). Comments on the paper, 'Volume compressibility of human abdominal skin'. *Journal of Biomechanics*, **12**, 481.

Walsh, R. J. (1964). Variation in the melanin content of the skin of New Guinea natives at different ages. *Journal of Investigative Dermatology*, **42**, 261–5.

Warndorff, J. A. (1971). The response of the sweat glands to β-adrenergic stimulation. *British Journal of Dermatology*, **86**, 282–5.

Weaver, J. A. & Stoll, A. M. (1969). Mathematical model of skin exposed to thermal radiation. *Aerospace Medicine*, **40**, 24–30.

Weinstein, G. S. & Frost, P. (1969). Cell proliferation kinetics in benign and malignant skin diseases in humans. *National Cancer Institute Monographs*, **30**, 209–46.

Weinstein, S. (1978). New methods for the *in vivo* assessment of skin smoothness and skin softness. *Journal of the Society of Cosmetic Chemists*, **29**, 99–115.

Whimster, I. W. (1965). An experimental approach to the problem of spottiness. *British Journal of Dermatology*, **77**, 397–420.

Whitton, J. T. & Everall, J. D. (1973). The thickness of the epidermis. *British Journal of Dermatology*, **89**, 467–76.

Wie, H., Engesaeter, L. B. & Beck, E. I. (1979). Effects of cyclophosphamide on mechanical properties of bone and skin in rats. *Acta Orthopaedica Scandinavica*, **50**, 629–34.

Wildnauer, R. H., Bothwell, J. W. & Douglass, A. B. (1971). Stratum corneum biomechanical properties. I. Influence of relative humidity on normal and extracted human stratum corneum. *Journal of Investigative Dermatology*, **56**, 72–8.

Wildnauer, R. H. & Kennedy, R. (1970). Transepidermal water loss of human newborns. *Journal of Investigative Dermatology*, **54**, 483–6.

Wilkes, G. L., Nguyen, A. & Wildnauer, R. (1973). Structure–property relations of human and neonatal rat stratum corneum. *Biochimica et Biophysica Acta*, **304**, 267–75.

Wilkes, G. L. & Wildnauer, R. H. (1973). Structure–property relationships of the stratum corneum of human and neonatal rat. II. Dynamic mechanical studies. *Biochimica et Biophysica Acta*, **304**, 276–89.

Wilkinson, D. S. (1961). Photodermatitis due to tetrachlorsalicylanilide. *British Journal of Dermatology*, **73**, 213–19.

Williamson, P. & Kligman, A. M. (1965). A new method for the quantitative investigation of cutaneous bacteria. *Journal of Investigative Dermatology*, **45**, 498–503.

Wilson, D., Goldman, R. F. & Molnar, G. W. (1976). Freezing temperature of finger skin. *Journal of Applied Physiology*, **41**, 551–8.

Winkleman, R. K. (1960). *Nerve Endings in Normal and Pathological Skin*. Charles C. Thomas, Springfield, Ill.

Winkleman, R. K. (1977). The Merkel cell system and a comparison between it and the neurosecretory or apud cell system. *Journal of Investigative Dermatology*, **69**, 41–56.

Witten, V. H., Sulzberger, M. B. & Wood, W. S. (1957). On the mode of action of alpha radiation from Polonium^{210} on human skin. *Dermatologica*, **115**, 661–70.

Wolf, J. (1937). Skin relief in man. *Bulletin interne de l'Académie des Sciences de Bohème*, **1**.

Wolf, J. (1939). Die innere Struktur der Zellen des Stratum desquamans der menschlichen Epidermis. *Zeitschrift für mikroskopisch-anatomische Forschung*, **46**, 170–202.

Wolf, J. A. & Maibach, H. I. (1974). Palmar eccrine sweating: the role of adrenergic and cholinergic mediators. *British Journal of Dermatology*, **91**, 439–46.

Wood, M. & Hale, H. W. (1972). The use of pigskin in the treatment of thermal burns. *American Journal of Surgery*, **124**, 720–3.

Wood, R. A. B., Williams, R. H. P. & Hughes, L. E. (1977). Foam elastomer dressing in the management of open granulating wounds: experience with 250 patients. *British Journal of Surgery*, **64**, 554–7.

Woodrough, R. E., Canti, G. & Watson, B. W. (1975). Electrical potential difference between basal cell carcinoma, benign inflammatory lesions and normal tissue. *British Journal of Dermatology*, **92**, 1–7.

Wrench, R. (1980). Epidermal thinning: evaluation of commercial corticosteroids. *Archives of Dermatology*, **267**, 7–24.

Wyburn-Mason, R. (1958). The reticulo-endothelial system. In *Growth and Tumour Formation*, p. 139. Kimpton, London.

Yamada, H. (1970). *Strength of Biological Materials*. Williams & Wilkins, Baltimore.

Yamamoto, T. & Yamamoto, Y. (1976). Electrical properties of the epidermal stratum corneum. *Medical and Biological Engineering*, **14**, 151–8.

Yamamoto, Y., Yamamoto, T., Ohta, S., Uehara, T., Tahara, S. & Ishizuka, Y. (1978). The measurement principle for evaluating the performance of drugs and cosmetics by skin impedance. *Medical and Biological Engineering and Computing*, **16**, 623–32.

Yannas, I. V. & Burke, J. F. (1980). Design of an artificial skin. I. Basic design principles. *Journal of Biomedical Materials Research*, **14**, 65–81.

Yen, A. & Braverman, I. M. (1976). Ultrastructure of the human dermal microcirculation: the horizontal plexus of the papillary dermis. *Journal of Investigative Dermatology*, **66**, 131–42.

Zackheim, H., Krobock, E. & Langs, L. (1964). Cutaneous neoplasms in the rat produced by Grenz ray and 80 kV X-ray. *Journal of Investigative Dermatology*, **43**, 519–34.

Zaias, N. (1980). *The Nail in Health and Disease*. Lancaster, MTP Press Spectrum Publications.

Pathological conditions referred to in the text

acne, 41
acrodynia, 148
actinic reticuloid, 163
aneurismal dilatation, 71
cancer, 66, 137, 144, 145, 159, 162, 164, 169
congestive heart failure, 33
cutis laxa, 100
dermatitis, 163, 165
dishydrotic eczema, 137–8
Down's syndrome, 15, 16
eczema, 19, 28, 60, 146, 147
Ehlers–Danlos syndrome, 99, 100
epidermal hypoplasia, 69
epidermolysis bullosa, 62, 83
erythema, 148, 159, 170
erythema multiforme, 148
exfoliative dermatitis, 148
fibrosarcoma, 169
haemochromatosis, 23
hyperpituitarism, 33
hyperthyroidism, 33
hypoalbuminaemia, 33
icthyosis, 60
lathyrism, 100
leprosy, 66
lichen planus, 19
lichen simplex chronica, 19

lichen ruber, 60
naevoid conditions, 18, 19
necrotising cellulitis, 182
onycholysis, 33
peripheral neuropathy, 33
photodermatitis, 163
photokeratitis, 159
pigmentation disorders, 170, 231
porphyria, 165
port wine lesions, 71, 167
proteoglycan storage disorders, 69
pseudo-xanthoma elasticum, 100
psoriasis, 13, 19, 38, 56, 57, 60, 149, 163
rheumatic fever, 20
scleroderma, 19, 83
sebaceous gland neoplasms, 164
seborrheic keratosis, 167, 168
solar urticaria, 163
squamous cell carcinoma, 169
summer prurigo, 163
tuberculosis, 20
typhoid fever, 20
urticaria, 148, 165
vasculitis, 70
Wilson's disease, 23
xeroderma pigmentosum, 23, 162

INDEX